Power Tool
Woodworking

by the same author:

Beginner's Guide to Woodwork
Beginner's Guide to Woodturning
Beginner's Guide to Furniture Making
Questions and Answers on Woodwork
The Powershop Manual
Woodworking Joints

Power Tool Woodworking

Gordon Warr

B.T. Batsford, London

Typeset by Servis Filmsetting Ltd, Manchester.
Printed in Great Britain by The Bath Press, Bath.

ISBN 0 7134 5720 1

B.T. Batsford Ltd
4 Fitzhardinge Street
London
W1H 0AH

Contents

Learning Resources Center
Santa Fe Community College
Santa Fe, New Mexico
(505) 471-8200

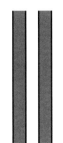

Acknowledgements

A very great deal of help and co-operation has been gratefully accepted from the following firms in the preparation of this book: AEG (UK) Ltd, Black and Decker, Robert Bosch Ltd, CeKa Works Ltd, Cintride Ltd, Clico (Sheffield) Tooling Ltd, Dunlop Powerbase, Elu Power Tools, English Abrasives Ltd, Florin Ltd, Garryson Ltd, Hitachi Power Tools Ltd, A. Levermore and Co. Ltd, Makita Electric (UK) Ltd, MK Electric Ltd, Meritcraft Ltd, Rabone Chesterman Ltd, Racal Safety Ltd, Record Marples Ltd, Skil (UK) Ltd, Trend Machinery and Cutting Tools Ltd, Wagner Spraytech (UK) Ltd, Wolfcraft.

My special thanks go to the following people who have been particularly generous in helping me: Geoff Brown, of BriMarc, John Costello, of Black and Decker, Jim Phillips, of Trend Machinery and Cutting Tools Ltd, John Roberts, of Robert Bosch Ltd, and Alan Wells, of Makita Electric Ltd.

In addition, my sincere gratitude goes to Ray Palmer, who transformed my pencil sketches into artwork in readiness for the printers, and to Mrs Gill Ferguson who typed the manuscript.

Last, but by no means least, my continued deep thanks go to my wife, Barbara, whose patience and understanding show no bounds to my lifelong passion for working with tools and creating with wood.

N.B. Many of the photographs of tools in this book have been taken with the guards raised or removed; this is solely to show the cutter or blade in the process of making the cut.

Guards, as appropriate to the tool and the operation being performed, must be used and be effective at all times for the maximum safety of the operator.

Introduction

In some ways, working with wood does not change as the years roll by. In other ways, though, very considerable developments have taken place since the 1950s. It's not the basic raw material which has changed – a prime piece of oak or pine is exactly the same today as it was a hundred or even a thousand years ago – but the way in which we fashion nature's most versatile of all raw materials.

Although timber responds so well to skilled hands and intelligent use, it does have its deficiencies. Man's ingenuity in overcoming some of these shortcomings, and also in exploiting still further the harvests of the forests, has led to considerable developments in the manufacture of board materials. We have had plywood for maybe three generations of craftsmen, but since World War II this product has been refined and improved, and grades and varieties extended. Chipboard is now used extensively, either in plain, veneered or decoratively faced form. There is hardboard in standard and decorative finishes, and while plastic laminates have now been in use for several decades, a more recent addition to board materials is medium density fibreboard, referred to as MDF for short.

Purists may well frown on man-made boards, which for the most part are wood substitutes, but we all use them from time to time, and therefore need to be familiar with their characteristics and uses, and especially their working properties. New materials open up new possibilities, and the world of woodworking has seen the introduction of many product developments which enable sheet material in particular to be used better. A wide range of hardware and fittings, many of them highly ingenious, have become available over recent decades, often designed so that man-made boards can be more effectively used. There are new adhesives, new polishes, new abrasives and improved screw technology. And there are even some new hand tools.

The power revolution

There is one area of woodworking that has seen more of a revolution than a change, and that is in the use of power. Not only has there been a considerable expansion in the use of static machines, but the machines themselves have also improved considerably in both range and performance. An even bigger impact than machines, though, has been the way workshops large and small have become equipped with portable power tools. From the smallest to the largest, there can be very few workshops indeed which do not have a portable power tool, and many have a good selection. Even shops well equipped with a comprehensive range of static machines are almost certain to have a selection of power tools in addition, as these can often perform certain jobs better than the fixed machines.

As well as use in the workshop, portable power tools by their very nature are ideal for the operator who works away from his base installing the products of his workshop. A wide assortment of power tools can be easily transported in the boot of a car, and so the portable power tools become the core of a portable workshop.

The first power tool to be produced and aimed at markets other than the giants of industry was the drill, introduced within a few years of the end of World War II. Early models were single-speed motors of limited power, and invariably with a chuck of small capacity. Other drills appeared, chuck size improved and power output increased, these last developments leading to greater boring capacity. The very considerable inventiveness of the manufacturers resulted in a wide range of attachments being produced for the drill. In their way these attachments were, and still are, excellent products, bringing to the small user operations that hitherto were not available – sanding and jigsawing to mention just a couple.

While many attachments work extremely well when using a drill as the power source, others have their limitations and are suitable for only very light work. For this and other reasons, such as greater demand and improved design and manufacturing methods, the small single-purpose unit with its integral motor appeared, usually very competitively priced when compared with equivalent functions as attachments.

The popularity of the self-powered tool led to new ones being introduced that were not normally available as attachments, examples being the planer and belt sander. Now there are yet more power tools appearing on the market – tackers, screwdrivers and biscuit-dowel jointers, to mention three that have appeared over more recent years in compact form and with price tags which appeal to a wide range of potential users.

Many of the power tools have been considerably refined since the early days of their introduction. Multiple and variable speeds have become standard features of many of the tools apart from drills, sanding frames have improved further the performance of belt sanders, and jigsaws are available with pendulum action, giving improved performance. The router is considered by many to be the most versatile of all the portable power tools, and now plunge action has become the standard form of this tool. Plunge action was first introduced in around 1950, but it took several decades to be fully appreciated over the fixed-body type. Router tables extend still further the usefulness of this tool, and the range of cutters or bits for use in the router is constantly being extended.

The introduction of tungsten carbide

Many power tools have benefited from the increasing use of tungsten carbide as a metal for cutting edges. Blades for circular saws are now readily available with TCT teeth, so too are blades for planes, bits for routers and even blades for jigsaws. Abrasives have also improved to keep abreast of their use on power tools, and methods of holding the abrasive sheet to the tool have likewise been developed.

What is the actual place of a portable power tool in a workshop whatever its size and nature? What are the advantages and limitations of this type of equipment? There is no single answer to these and similar questions relating to any equipment in any workshop. They will vary from user to user, workshop to workshop, and from one branch of woodworking to another. The needs of the furniture-maker will be different from those of a boat-builder. The requirements in a workshop with a wide range of static machines will not necessarily be the same as in a workshop with minimal fixed machinery, and what

the professional user expects and needs from a power tool will be distinct from the requirements of the occasional user.

Even in workshops of moderate size, a portable power tool may, in certain circumstances, be used very successfully as an alternative to a static machine. A belt sander, for instance, is in many ways more useful, at a fraction of the cost, than a fixed sander of the overhead pad type – it can be used on an assembled carcase, for instance, which a fixed machine cannot normally tackle. Often portable power tools have a very useful role to play in supplementing the machine alternatives to specific power tools, and the average professional shop equipped with a spindle moulder will be sure to find plenty of use for a router. In fact, some portable power tools can carry out certain operations that most static machines cannot, for example moulding the centre of a panel with a router, and making internal cuts with a jig or sabre saw.

In general terms, portable power tools are not substitutes for fixed machines but complement them. This is particularly so in cases where the power tool can more readily carry out the function than the machine, for instance using a power planer to form a rebate. Rebating can be carried out on some planing machines, but if this is not possible the power plane offers a simpler alternative to rebating on the spindle moulder, assuming one is available. Power tools score heavily in terms of economy, not just in cash terms, but also in relation to space requirements. Few workshops have space to spare, so often the acquisition of a portable power tool provides a means of carrying out an operation – and often a series of operations – when it would be difficult to accommodate even a smallish static machine.

The total number of power tools available from the several manufacturers who produce this kind of product is now quite vast. Though there might appear to be a certain amount of duplication of product between one maker and another, there are often differences that are not at first obvious. Some manufacturers are strong on certain products, but don't necessarily produce every type of power tool currently available. Certain manufacturers offer a wide range of power tool accessories, such as a router table, others have a limited range. Often an accessory from one manufacturer will accept products from another maker – a drill stand, for instance, provided it has the European standard fitting for a 43mm collar, will accept any make of drill with a collar of this size.

The different ranges

Many, but not all, of the power tool manufacturers make their products in two ranges, the professional range, and the amateur models. The latter are also known as the consumer or household product. Because there are often two or three models of each type of tool within both the ranges, the difference between the top of the amateur range and the lower end of the professional models is often very small. On the other hand, though, there is quite naturally a considerable difference between tools from the economy end of the consumer products and the best of the professional range.

Often a better quality amateur tool will be adequate for the needs of the professional user where the frequency of use of such a tool is likely to be limited – a jigsaw, for instance, might only be used on fairly rare occasions by a cabinet-maker, and even then for cutting material that is invariably quite thin. On the other hand, the joinery contractor who carries out a lot of work on site will probably want a portable circular saw of large capacity and rugged construction. Tools of a given capacity will perform well up to and including that capacity, but not beyond. Any overloading of the tool will impose both a mechanical and electrical strain.

All the products from the established manufacturers offer very good value indeed, and therefore represent a sound investment which should give excellent service for all users, and for the professional the products in most cases will pay for themselves in a short space of time. It is always wise, though, to obtain the product that will cope adequately with the work intended for it in terms of both accuracy and capacity. It cannot be expected, for example, that an economically-priced light router will be able to carry out the same heavy cutting that a powerful top-of-the-range professional tool can tackle. Wherever possible, therefore, for both professional and amateur use, it is always wise to obtain the power tool that will cope with the workload intended with some capacity to spare.

1 ‖ Safety

We have all become far more safety-conscious over recent years, and there is much legislation to guide both the manufacturers and the users so that accidents are reduced to a minimum and injuries kept to only minor ones. It would be wishful thinking to believe that every form of accident and injury could be eliminated, but what we should realize is that an extremely high percentage of them are avoidable. Safety is basically a combination of good design on the part of the manufacturer and an understanding in the fullest sense of the equipment on the part of the user combined with an attitude of mind. Nobody seeks to do harm to themselves, or indeed others, but when equipment is used in ignorance, or when sound practices are overlooked in favour of expediency, then the risks of something going wrong rise considerably.

Manufacturers of portable power tools are now obliged to ensure that their products conform to high standards of safety in order to have them approved. This doesn't just mean that the tool has to be sufficiently rugged so that it can cope physically with the work it is expected to carry out, but that all appropriate guarding is provided, the electrics and the insulation are of the highest standards, and that, where applicable, dust bags are provided. Manufacturers are very anxious to ensure that their products enjoy a good safety record, and most go beyond just the minimum requirements. As well as including instructions for use with the products, basic rules of safety relating to that tool are now almost always provided. However familiar a person is in general terms with powered equipment, be it machines or portable power tools, it is essential to read the literature relating to any new product not used previously.

Safety regulations do vary a little from country to country, and the manufacturer ensures that his products satisfy marketing requirements. In addition to the safety requirements of the products themselves, there is legislation relating to their use, and to operator safety. In Great Britain this is covered by the Factory Acts relating to the safe use of woodworking machinery, although there is some uncertainty in the regulations as to whether all portable power tools are classed as woodworking machines. The regulations are primarily designed for the protection of the employee, and while they are not a legal requirement for the amateur working at home, or the self-employed person who operates a one-man business, they obviously contain much sound advice applicable to any user.

General safety rules

There is little point in listing here all the do's and don'ts relating to every portable power tool. The list would be very long and repetitive of the guidance given for the individual tools. Reference to safe working practices is given in the relevant chapters. There are, however, certain points of safety that are common to all power tools and these are stressed here.

1. Become familiar with the tool, and ensure that all guarding appropriate to the tool is in use according to the mode of operation.

2. Do not overload the tool; keep within stated capacities. Ensure that all cutters and blades are sharp, correctly matched to the tool and properly secured in place. Electrically isolate the machine when changing cutters and blades, or when carrying out adjustments to these parts.

3. Keep the working area clean and tidy. Keep trailing leads to a minimum, and be aware of the danger of them tripping the feet. Provide adequate lighting.

4. See that the workpiece is properly secured, and that any temporary supports to the wood being worked on are adequate. Avoid over-reaching, and ensure that the stance taken is comfortable.

5. Avoid loose cuffs, ties, long hair, bracelets and similar items which are potential sources of danger.

6. Never let children, nor indeed uninitiated adults, use the tools. Keep any onlookers at a safe distance.

7. Never leave a piece of equipment running in the absence of the operator. Isolate the tool if there is any possibility of interference.

8. Wear personal protection as appropriate to the tool and the operation. This is discussed a little later (see pages 17 and 18).

Electrical safety

The vast majority of mains-powered portable tools currently being manufactured are double insulated. For a tool to be classed as double insulated in this way, the body of the tool must be made of either plastic or metal with the electrical components entirely contained in a plastic housing within the body. Various plastics are used for the bodies. Often these are reinforced but an essential characteristic is that they remain unaffected by the heat generated when the tool is in use. The materials used on early all-plastic power tools were not sufficiently resistant to the effects of heat, with the result that bearing housings in particular soon showed signs of allowing movement of the main drive spindle. Plastic bodies have a high resistance to impact, although every care must be taken not to drop the tools or cause other physical abuse.

Double insulated tools do not require an earthing connection back to the mains supply. They do, however, always need connecting to a suitable power outlet, this to be via a fused plug for further protection. Power tools must never be connected to a lighting supply, even tools of low wattage. Under certain conditions of usage power tools can draw a very much higher wattage than their rating, with consequent risk of damage to the lighting supply which is not intended for a high wattage. Some older tools, and a limited number of current tools as well, are not double insulated. These are known as earth products and require earthing via a three-core lead back to an earthed supply. This wiring arrangement should never be altered in any way, and the earthing provision should be maintained at all times. Maintenance of the tool should include inspection of the lead and the plug, and particular attention should be paid to ensuring that the anchor clip of the cable within the plug is effective. The tools should never be held by their leads, and no physical strain should ever be imposed on the lead or the connections of the lead. Moulded-on plugs are becoming increasingly common, and these should not be tampered with in any way.

An extension lead is often required, particularly when power tools are used away from the workshop.

Always ensure that such a lead is capable of carrying the load demanded by the power tool. A coiled lead with a heavy load passing through it is likely to heat up and should always be unwound before use. This precaution is particularly important when using tools of a high wattage rating such as large circular saws, or where more than one piece of equipment is being operated at the same time on an extension lead. Some leads incorporate a thermal cut-out as a precaution against possible overheating.

Mains-operated power tools should never be used out of doors in anything other than dry weather. Every care must be taken to ensure that all leads, and especially plugs and sockets on any extension leads used, are kept clear of water or any damp areas of ground. The best precaution when using portable power tools out of doors, or in any situation where there is some physical risk to the power supply, is to use a residual current-operated circuit-breaker, known as RCBs.

Residual Circuit-Breakers

An RCB is a protective device designed to isolate the electrical supply when a fault on the circuit causes a

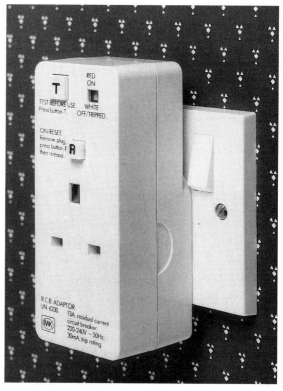

1 *A residual circuit breaker with test and reset buttons*

current that exceeds a pre-set level to flow to earth. The commonest causes of an earth flow current are the breakdown of the insulation as a result of deterioration, physical damage to the lead or ingress of water. Earth fault currents can cause fire, as well as potentially fatal shock and injury to the operator. For protection of an individual circuit the RCB is simply plugged into the socket, and the power tool lead is then plugged into the RCB. The RCB incorporates a test button that simulates an earth leakage when pressed and causes the breaker to trip.

The reset button on the breaker brings the trip device back into operation, but only if all is well with the circuit. On many RCBs the reset button is located on the inner face. This is an additional safety feature as access to the button can only be gained by removal of the breaker from the socket. A suitable rating for an RCB for use with a portable power tool is 30 milliamps. The common rating of 10mA, although more sensitive to earth leakage, is not considered ideal for power tool use; it can be tripped by an abnormality in the balance in the electrical flow caused by the switching mechanisms of some cheap products. This is in no way meant to suggest that cheaper tools are in some way inadequate for the work they are intended to perform. The switching mechanism of power tools is a complex part of the design, and the quality of this is related to the price bracket of the particular product. The 30mA breaker provides adequate protection without tripping for the wrong reasons.

There is a common belief that an RCB will give total protection against electrical shock should the body be exposed to live wires. Though this is not so, the RCB responds so quickly to such a situation that the supply is tripped almost instantaneously, and the flow of current is of such a minute duration that no harm is done.

Eye protection

However much the moving parts of a power tool are physically protected and even if a system of debris collection by vacuum is installed, there is always likely to be dust in the air when wood is being worked. Apart from the irritation that the dust can cause to our eyes, there is some risk of a larger particle becoming embedded in the eye. At the very least this will cause both disruption and discomfort but can be easily avoided by the use of appropriate eye protection.

Safety glasses and goggles are not expensive, and certainly not when compared with the irreplaceable value of our eyesight. Goggles are available with either twin lenses, or a single panoramic lens that

restricts the field of vision very little indeed. The frames are usually made of flexible PVC, with lenses of clear acetate, polycarbonate or Panomist, the last having anti-misting properties. Goggles provide complete all round cover to the eyes, and because of this they are provided with vents. Goggles give the best protection when the dust level is fairly high, or when the eyes are particularly sensitive.

An alternative to goggles is eye shields. These provide protection from the front, but do not totally enclose the eyes as goggles do. Most models have either clear or green anti-glare polycarbonate lenses, and these are usually replaceable. One form of eye shield is a lightweight one-piece version that gives wide angle vision and can be worn over ordinary spectacles.

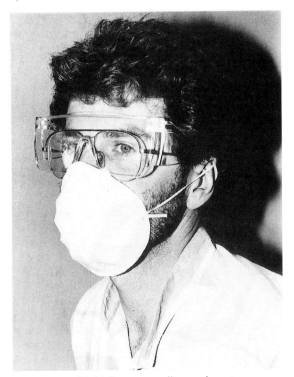

2 *This eye shield fits over ordinary glasses*

A third variety of eye protection is safety spectacles, which look very much like ordinary spectacles. These are produced with either clear or tinted lenses, and also with clear, removable side shields. For the person who normally wears glasses, it is possible to obtain lenses prescribed for vision correction in both toughened glass and polycarbonate.

Nose protection

The smell of freshly-worked pine is a delight to all, not just the craftsman. The dust of other woods, though, can cause severe nasal irritation, and can bring on bouts of sneezing that persist for hours. Working woods such as mansonia, ebony and makore, especially when being sanded during turning, can produce a dust that quickly affects the breathing system, often acutely. Different people react to wood dust in different ways. Some in very adverse conditions can become mildly ill. What we do not know is what is happening within our own bodies, and particularly what the cumulative effect of frequently inhaling such irritants is. Even those timbers with pleasant smells that do not create dust that we consider obectionable can nevertheless represent a danger to our health. Breathing in any dust-laden atmosphere, however sweet the odour, cannot do any of us any good. Those with chest ailments or respiratory problems, such as asthma, must be specially careful regarding dusty atmospheres as they are obviously more susceptible to the hazards involved. Everyone who works where there is doubt regarding the cleanliness of the air should take all reasonable precautions to safeguard their health.

Dusts masks are relatively cheap, although the cheaper models have a fairly limited life of a few hours, after which they should be scrapped. Not all masks give protection against sparks, while better types have improved filtering qualities giving better protection. More expensive masks often include non-return valves to make breathing easier.

Masks should only be used within their prescribed limits of performance. They lose their efficiency if used for excessive periods of time, and must not be washed as this destroys their filtering properties. While better quality masks reduce odours, they do not purify the air, nor are they intended to be used where there are toxic fumes present.

Dust collection

Often the problems of dust can be reduced at source. Some types of power tool are fitted with dust bags. These should be maintained in good condition along with the fans and airways that are a part of the dust collection system of the tool. With certain models of power tool the dust bag is an optional extra, but it is always a worthwhile facility for a fairly small additional cost. However, some tools produce chippings rather than dust, and while it is not practical to have a

3 *Personal protection for operator safety*

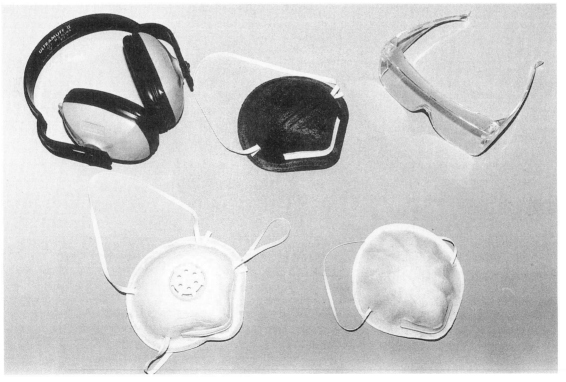

chippings bag attached to a hand-held router, they can be fitted to suitable planers. The difficulty lies in the fact that chippings can be generated at a very fast rate and the bag is soon filled.

The problem of debris collection can be overcome by vacuum extractors, and at least two of the portable power tool manufacturers offer small models particularly suited to a variety of power tools. The tool must have provision for dust collection, and in many cases adaptors or connector spouts are required to make the necessary link-up. The vacuum units are mounted on castors for ease of mobility and have a flexible hose for connecting to the power tool. Accessories are available so that these extractors can be used as normal vacuum cleaners and become of additional use for maintaining clean conditions in the area of work. Vacuum extraction units of this type are particularly useful where power tools are being used 'on-site' during alteration and maintenance work and the need to keep the nuisance of dust and debris to a minimum is especially important. The Elu vacuum cleaners incorporate an ingenious switching arrangement. The tool is plugged into the cleaner and the cleaner into the mains. Switching on the tool automatically brings the cleaner into operation, with the alternative of independent use and control.

4 *Vacuum dust collection with automatic cut-in*

Ear protection

Most powered means of working wood, whether by machine or tool, tend to be noisy, and some more than others. Sawing of hard woods and routing work probably create more noise than most other operations. While the risk to the ears might not be severe, exposing the aural system to noise of even moderate levels can have a cumulative effect. The efficiency of our ears can slowly diminish when exposed to undesirable noise over extended periods, and the offending noise does not have to be extreme for loss of hearing to occur.

The answer is to use ear muffs. Modern muffs are comfortable, light in weight, and very efficient. Like all types of safety equipment, they have to reach high levels of efficiency, and undergo thorough testing before being approved by the appropriate safety bodies. For use in connection with woodworking activities, those classed as being of general purpose attenuation are considered suitable.

2 ‖ Drills

It was the drill that started the power tool revolution, and it was from this that virtually all the present vast range of power tools have developed. The drill itself has, of course, undergone many changes and almost endless improvements over recent decades, making the present day generation of drills very different products in comparison to the earlier models.

Motors

The motors fitted to most power tools are of the series-wound universal type, their size and power being proportional to both the chuck capacity and the drill's rating to bore holes of given diameters in the three commonest materials in which drilling or boring usually takes place – wood, metal and masonry. Motor size is rated in watts, but what is stated in the description of the drill in the manufacturers' brochures is invariably the 'input' wattage. The effective output power of the motor in watts is around 60 per cent on average of the input wattage. Thus a drill of 650 watts will have an output wattage of approximately 390 watts.

One of the commonest causes of drill breakdown is motor failure. This is usually a result of misuse, most often through overloading. The wire used on the armatures, along with the insulation, the commutator and brushes, and the main bearings at either end of the motor spindle, are all designed in relation to the drilling rating of the tool. Using the drill beyond its capacity, or abusing it by expecting far too high a feed rate, will very likely lead to failure. The motor becomes overheated and the insulation can fail, resulting in the motor being burnt-out. If for any reason a drill should stall during use, the power must be switched off immediately, or overheating and burn-out will result. Sensibly and moderately used, an electric motor has a very long life apart from renewal of the carbon brushes. Drills can also break down or fail to operate at their best because of excessive wear in the gears or the bearings, or because the switch ceases to operate as it should. Switches on power tools and drills in particular are more complex than might be thought, and the switch is a major factor in drills of different size, quality and price.

What has been said about the wattages of motors, and the dangers of overloading, of course applies to all power tools. Indeed, most power tool manufacturers use the same size and type of motor across a range of their products, so that the number of different motors they need to produce is quite small in comparison with their total number of products.

Chucks

Nowadays, the vast majority of drills are fitted with either a 10mm ($\frac{3}{8}$in) or 13mm ($\frac{1}{2}$in) chuck. The 6mm ($\frac{1}{4}$in) chucks of the early models have now been superseded as the drills themselves have been upgraded. What is not always realized is that most patterns of chucks as fitted to power drills also have a minimum holding capacity. This is usually 1.5mm ($\frac{1}{16}$in). As well as the differences in size, the quality of the chucks fitted also varies, broadly falling into the three categories of light, medium and heavy. The power tool manufacturer has to be relied upon to fit a chuck of appropriate duty-rating to match the capacity and performance of the drill. This is another factor differentiating drills that superficially look the same but are in different price brackets.

The usual way of mounting the chuck is by threaded connection to the spindle that links up with the gearbox of the drill. Virtually all manufacturers now design their drills so that an internal, or female, thread is required on the chuck, and two sizes and thread forms have become almost universally accepted. These are $\frac{3}{8}$in $\times$ 24 (threads per inch) and $\frac{1}{2}$in $\times$ 20. Note that it is the imperial diameter which has been accepted. There is no direct metric equivalent. Chucks are therefore interchangeable providing the diameters correspond, but it would be foolish to replace a 10mm ($\frac{3}{8}$in) capacity chuck with one of 13mm ($\frac{1}{2}$in) capacity in the belief that the drill was being uprated. It should also be noted that the diameter of the threaded connection between chuck and drill is not directly related to the capacity of the chuck.

Many drills have reversing action, and this requires an additional feature to the chuck. As the thread on the chuck is always of normal right-hand form, there is clearly a strong possibility of the chuck unscrewing itself off the spindle when used in reverse. This problem is overcome by fitting a lock-screw between the chuck and the spindle, the screw being inserted through the inside of the chuck into the centre of the spindle. To gain the locking effect, this screw is of left-hand form, and must first be withdrawn before removing the chuck.

The chuck must be unscrewed anti-clockwise for removal. It will normally be quite tight on the spindle, and to slacken it the chuck key is inserted and given a smart blow. As the resistance to the force of the blow is taken by the gears, some care is needed and excess force should be avoided. On many makes of drill, there is provision between the chuck and the body for the spindle to be gripped by a spanner. With these drills, the spanner should be used in order to keep any strain (caused by removing the chuck) from the working parts of the drill.

Collar of drill

Happily, there is now standardization on the size of the neck, or collar, by which the drill is gripped when used in a drill stand, or when attachments are added.

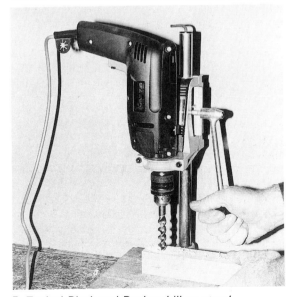

5 *Typical Black and Decker drill on stand*

For the vast majority of drills, this is 43mm, although certain heavy duty industrial drills have a neck diameter of 57mm. Many manufacturers have taken advantage of the collar by using it to mount a side handle. This clamps on to the collar, with the handle itself being rotated to tighten the assembly. This is a very satisfactory way of providing a second handle, as it can be rotated through 360 degrees and locked in any position for convenience and comfort. In turn, the side handle provides an easy means of incorporating a depth stop, a useful feature for many boring operations. The depth stop may be locked in position either by rotating the handle, or by using a separate thumb screw if provided. The alternative way of mounting a handle is by threaded connection into the body. A boss is moulded into each side of the body casing. These have an internal thread and the handle screws into one of these. The choice of handle position in this case is therefore down to two, but of course a collar-mounted handle can be used on any drill with a 43mm (1⅝in) collar, even if this type of handle is not provided as standard.

Features

The range of speed options, and speed control, is very varied. Many models of drills combine mechanical means of control, and the more advanced tools incorporate sophisticated electronics. As many drills offer hammer actions, reversing facility and torque control, the combination of features to be found on many of the better tools is considerable, and this accounts for the wide range of drills offered by most of the leading manufacturers.

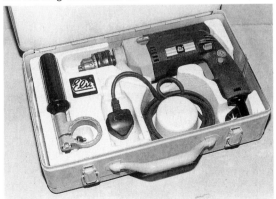

6 *The Elu variable-speed reversing percussion drill*

Very few drills offering simply single speed operation and no more are now produced, and these are usually at the economy end of the manufacturers' ranges. Such a single speed is likely to be in the region of 1,800 rpm, as it has to be a compromise to cope with as many drilling, and other, operations as possible. Another factor influencing the manufacturers' choice of speed is the tool's rated drilling capacities. Small size bits will function quite well at a moderately high speed – the larger the bit being used the slower the speed needs to be. The speed for boring wood should normally be rather faster than that needed when drilling metal, which is one reason why the single speed drill lacks the flexibility of those drills with dual, multiple or variable speed.

7 *Peugeot drill in AEG stand*

With variable speed drills, the trigger acts as an accelerator. Thus the speed is infinitely variable from zero up to the maximum, but there is no way of assessing the rpm other than by judgement. Pressing the trigger controls a rheostat; this governs the flow of electricity to the motor, which is designed to run at full speed at maximum electrical input, and at lower speeds according to the electrical flow. However, the motor only delivers its maximum power at full speed. This means that at anything less than maximum rpm, the power is proportionally reduced. Clearly though, variable speed has many advantages, and is particularly useful at the start of boring, and also at the point of breakthrough on through holes.

All drills, except those with electronic circuitry when used sensibly, have what is known as a 'no-

load' speed, and a speed 'under load'. This means that as the boring imposes a resistance on the motor, therefore acting like a brake, the result is a slowing down of the motor. The slowing down effect, which gives the speed under load, will depend on a number of factors, including the size of hole being bored, the hardness of the material in which the hole is being made, the sharpness of the bit, the pressure exerted by the user on the drill, and the quality and rating of the power tool. Excessive slowing down for whatever reason leads to both mechanical strain and overheating. Prolonged overheating leads to motor failure.

8 *Heavy-duty Black and Decker drill*

Two-speed drills have different predetermined upper and lower speeds depending on the model, typical speeds being around 2,000 rpm and 3,000 rpm. The two speeds available are provided mechanically via gearing that links the motor to the spindle that carries the chuck. The speed selection is made by a small lever either on the side or top of the body, and this should only be activated when the motor is stationary. On the basic two-speed models, the motor runs at a constant no-load speed, and the gearbox provides the only means of varying the speed of the chuck.

Some manufacturers have taken the above idea a stage further by combining the mechanical gearbox with variable speed on the trigger. Thus the trigger controls the speed either from zero up to the lower limit, or by moving the gear lever from zero up to the upper limit. Again with these models, the trigger controls a rheostat, with the result that lower speeds throughout both speed ranges suffer from power loss.

The move over recent years has been towards having electronic speed control. The rheostat is replaced by electronic circuitry. The essence of this is that the load on the bit is detected and the flow of electrical power to the motor is adjusted accordingly. Thus the speed selected for a particular operation is maintained throughout regardless of varying loads and without the fall-off otherwise experienced. Drills are now being produced with eight and even more pre-set maximum speeds selected by rotating a small wheel, which is part of the trigger. These are combined with a two-speed mechanical gearbox to double the maxima, and all variables from zero up to the top speed decided upon. The flexibility of these drills in terms of speed is therefore very considerable, a higher speed of up to 3,500 rpm, and up to 18 preselected top speeds up to the maximum, with total control throughout from zero upwards.

With the introduction of tungsten carbide tips, power drills soon proved their worth for drilling masonry and similar very hard and abrasive materials. With the correct combination of bit and tool, holes can be produced accurately and quickly in a wide variety of building materials. For the operation to be effective, the power drill needs to have 'hammer-action' facility. Hammer-action is when the chuck and bit are being moved forwards and backwards as they rotate. The amount of axial movement is not great, around 2mm ($\frac{3}{32}$in) on most drills. The hammer-action is created by a ratchet system adjoining the gearbox. A hardened steel disc with a series of raised notches causes the bit to vibrate at a very high rate, anything up to around 48,000 blows per minute when not under load.

Because hammer-action is only suitable for materials composed of grit-like particles, all but specialist heavy duty masonry tools are dual-action, that is hammer or normal drilling. Hammer-action is brought in by operating a lever on the upper part of the body of the drill, although on some older models a ring behind the chuck is partially turned to engage the ratchet mechanism. Drilling masonry can be carried out by using non-hammer power drills, but a slow speed must be used in order to prevent rapid blunting of the bit.

Most materials for which masonry bits are used are both hard and abrasive, and the dust created during drilling consists of fine particles with the same characteristics. Such dust can cause damage if allowed to enter the drill through any air vent holes, and can also cause problems to the chuck if there is constant ingress of the drilling debris while the drill is in use. It is wise to use a baffle over the bit that will collect the bulk of the waste, and regular cleaning to brush away all signs of dust helps to minimize possible problems.

Some woodworkers will hardly ever need to drill into masonry, others will have frequent need to drill into building structures in order to secure the products of their workshops. For those who carry out a lot of on-site work, a drill is now produced that combines the best of electric drill technology with that of pneumatic drills. Usually called the electro-pneumatic rotary hammer drill, the percussion action is created by air pressure generated by the motor. This pattern of drill delivers around 4,500 blows per minute, and is far more efficient than the normal hammer-action drill. Whereas the hammer-action drill requires pressure from the operator in order for the bit to penetrate, the electro-pneumatic drill delivers blows that enable the bit to advance in the hole and the operator need only hold the drill in place and apply relatively little pressure. They are highly effective, and far less tiring, and the makers who produce this kind of drill claim that they are three times faster than hammer-action, with only a third of the pressure being required.

Electro-pneumatic drills require a special type of masonry bit, known as SDS bits. These are tungsten-

9 *Bosch electro-pneumatic percussion drill*

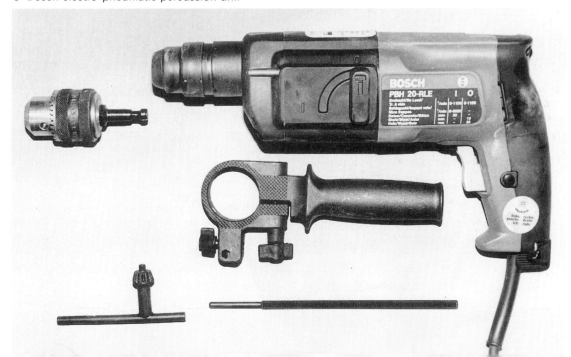

carbide-tipped, but are ground rather differently from the usual pattern of masonry bit, as they cut more by the blow effect than the rotational one. They also have a special shank which fits directly into the drill without the usual three-jaw chuck. They are slipped in and out of place simply by moving a sliding collar, and a special chuck adaptor is used for mounting a standard pattern of chuck to enable normal drilling to be carried out. Because the primary function of these drills is drilling in masonry and similar materials, the maximum rpm is around 1,200 or less. They have variable speed up to this limit, with some models having reverse facility.

Power screwing

Torque control is another development in power tool technology and is a feature that applies particularly to driving in screws. While in theory any drill with a screwdriver bit in the chuck will drive in screws, in practice the method is not too satisfactory in most instances. With most of the older generation of fixed-speed drills, including two-speed models, the speed is too high and it is difficult to control the screwing, particularly when the screw reaches the end of its intended drive. Overdriving, and subsequent damage, are difficult to avoid. Variable-speed drills are better but still not entirely satisfactory because of the power/speed problem of non-electronic control.

Torque-control drills have a clutch-type mechanism that cuts out the rotation of the chuck when the

10 *Black and Decker screw drill*

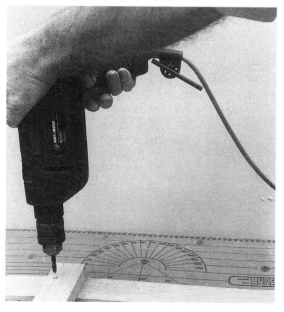

load on the bit reaches a pre-set limit. The cut-out point can be adjusted to provide multiple settings according to the length of the screw, the hardness of the wood, and the amount of preparation given by way of boring pilot holes. By adjusting the setting of the torque control, the penetration of the screw into the wood can be governed so that the head of the screw finishes up at a predetermined position relative to the surface of the wood.

Screwdriver bits are usually provided with drills designed to have screwdriving facilities. The bits may be double-ended, suiting cross-head or slot-head screws, or single-ended, and different sizes of bits are available to suit various gauges of screw. The bits incorporate a hexagonal section to their shank so that a positive grip may be gained when secured in the chuck.

While power screwing operates well on both slot-head and cross-head screws, it is at its best on the latter because of the better engagement of the bit in the head of the screw. This minimizes the risk of the bit slipping off the screw, with possible damage to the wood resulting.

Variations

Occasionally there is a need to drill holes in confined spaces, where a standard pattern of power drill, even if a fairly small model, cannot gain access. The angle drill is intended for use in such restricted conditions of working. Because the chuck is at right angles to the body, which is itself quite slender, the amount of clearance required for operating the drill is small.

Bodies and handles

The vast majority of drills are of the pistol-grip pattern, with the switch of the trigger type. Drills normally have provision for locking-on the switch, a desirable feature when, for instance, the drill is being used in a drill stand. While the pistol-grip pattern of handle is comfortable and convenient for the vast majority of uses for which the drill is employed, it does mean that the pressure exerted on the drill is not in line with the axis of the bit. For particularly heavy duty drilling, some makers produce drills with the handle at the end, which enables rather more pressure to be applied during drilling. These are usually referred to as D-handles, and are only likely to be found on drills rated at around 650 watts and above.

Certain older types of plastic used in some early-model plastic-bodied power tools did not prove to be

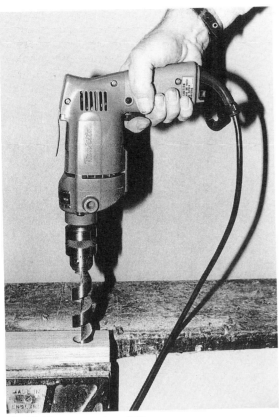

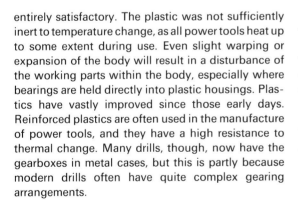

11 *Makita percussion drill*

12 *Black and Decker 'Plus' drill*

entirely satisfactory. The plastic was not sufficiently inert to temperature change, as all power tools heat up to some extent during use. Even slight warping or expansion of the body will result in a disturbance of the working parts within the body, especially where bearings are held directly into plastic housings. Plastics have vastly improved since those early days. Reinforced plastics are often used in the manufacture of power tools, and they have a high resistance to thermal change. Many drills, though, now have the gearboxes in metal cases, but this is partly because modern drills often have quite complex gearing arrangements.

Drilling capacities

The various manufacturers of power tools issue brochures and other promotional literature in which the particular features of their products are listed, and their working capacities stated. The latter in the case of drills is usually stated as maximum diameters of holes which can be made in various materials, nor-mally masonry, steel and wood. While mild steel is fairly consistent in its composition and hardness, masonry, concrete and brick are highly variable in their hardness and therefore in their resistance to drilling. This is equally true of wood, and as well as the denseness of the particular piece, the presence of hard knots will affect the drilling, as will the type of bit used and the direction of boring relative to the grain. Various patterns of bits are discussed in Chapter 3. Any bit used in end grain will find the wood offers greater resistance to being cut than in cross-grain working, this is because of the way in which the fibres of the grain have to be severed.

Another factor that affects the boring capacity in wood is the depth of the hole to be made. The deeper the hole, the greater the resistance to the boring; this is largely because of friction with the walls of the holes and the chippings that have to be ejected. Shallow holes, or holes which pass all the way through fairly thin material, generally require less energy for their making.

All the above means that the manufacturers' recom-mended drilling capacities must be taken as a guide only, although clearly those drills with greater wattage ratings will tackle larger holes than the lower powered ones. The important points to remember at all times and with all power tools is to use them with sensitivity, avoid overloading, and regard any excessive heat build-up in the tool as a severe warning signal. Remember, too, that with both steel and masonry it is often feasible and good practice initially to drill a hole smaller than that actually required, then open this up with a larger bit. This considerably reduces the load on the motor, but the vast majority of bits for wood do not lend themselves to this procedure.

3 | Boring and Drilling Tools

Although the power drill can be used to carry out a wide range of operations, its prime function when first introduced was to bore holes, and this has remained so throughout its development. In its early days, no bits were produced specially for the portable power drill. Bits for machine boring were available but most of these had 13mm ($\frac{1}{2}$in) shanks, which made them unsuitable for most of the older drills, which had chucks too small for the existing machine bits.

Twist drills

One bit which has remained forever popular for use in the power drill is the engineer's pattern of twist drill, sometimes known as the morse drill. These drills are produced in a great range of types and sizes, the bulk

13 *Various series of morse drills*

of which are intended for drilling various metals. Long before the power drill became a part of so many workshops, however, it was discovered that the standard type of twist drill would bore holes in wood, and most woodworkers had a selection of these in their kits. Although they were not intended to be used in a hand brace with its 'alligator' pattern jaws, they were perfectly satisfactory in hand drills, or wheel braces as they are often referred to. Indeed, the engineer's twist drill became very much a dual purpose drill, as there are very few if any woodworkers who do not have the need to drill metal from time to time.

Originally, the bulk of twist drills were made of carbon steel. These were of limited hardness and soon lost their cutting edges on metals unless these were fairly soft or mild. Once the capacity to cut has been lost, the cutting edges do little more than rub the work,

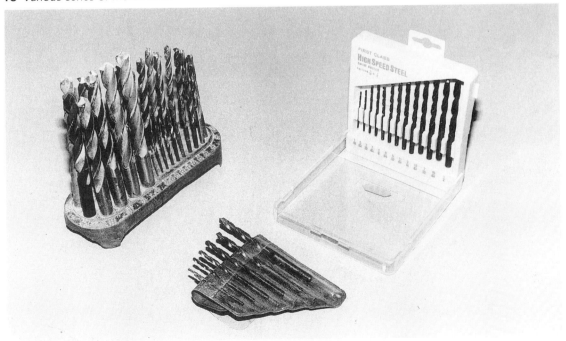

causing friction. Friction generates heat, the heat affects the molecular structure of the drill, and its tip is softened. The colour changes to blueish grey, and this is usually known as burning. The affected tip of the drill, often extending back 13mm ($\frac{1}{2}$in) or more, has to be ground away until unaffected steel is reached.

The pattern of dulling of cutting tools takes place on all those produced for wood, regardless of the metals used for the cutting edges. Once the efficiency of the cutting edges has been lost, rubbing follows, which quickly reduces any cutting capacity to a very low level. This results in overheating and scorching, a very slow rate of cutting, and a poor surface to the workpiece. Excessive pressure of the tool on the work to try to compensate for dull edges makes a poor situation even worse, and in addition overworks the motor, which in turn can lead to permanent damage. This is why all cutting tools, including those for boring holes, must be maintained in a sharp condition.

Because of the problems with burning affecting twist drills made of carbon steel, most drills currently produced are of high speed steel (HSS). This is a much harder steel than carbon, retains its cutting edge for a longer period of time and does not change its characteristics if overheated in use. Carbon steel twist bits, although suitable for wood and cheaper than HSS, do not represent good value except for very limited use. HSS are to be preferred as they will give a

14 *Lip and spur bits*

much longer life and can be used for a wide range of materials.

The bulk of twist drills whether of HSS or carbon are of the pattern known as jobbers. This term means that they are of a standard length related to their diameter and have parallel shanks. The spiral or helix is of medium pitch, and the point has an included angle of 120 degrees. Other types include those with the shank stepped down to 6mm ($\frac{1}{4}$in), a 'long series', 'quick' and 'slow' spirals, tungsten tipped and those with morse taper shanks.

For power tool use in relation to wood, one of the big advantages of the twist drill is the wide range of sizes they are produced in, making them particularly suitable for the preparation of screw holes. They are made in fractional imperial diameters from $\frac{1}{64}$in to $\frac{1}{2}$in and above, and in metric from 0.25mm to 13mm and larger. In addition, mostly for engineering use, letter sizes from A to Z, and number sizes from 1 to 80, are also produced.

HSS twist drills have a very long life when used exclusively for wood, and can be sharpened in a number of ways. They can be ground freehand on a suitable grinding wheel, but considerable skill is needed for this in order to retain the original angles, and keep cutting edges of equal length. If they are ground even slightly eccentric they will always bore oversize. While for most woodworking purposes this small inaccuracy will probably not matter, it becomes more critical on holes made in metal. It is possible to

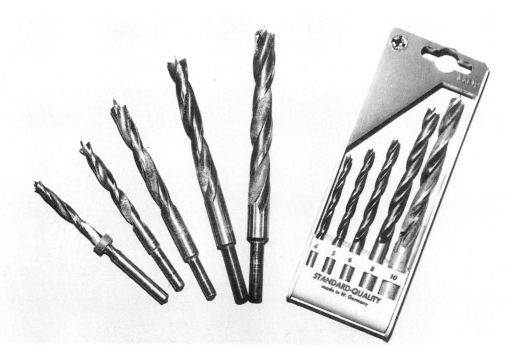

obtain attachments that hold the drill at the correct compound angles against the grinding wheel, and drill sharpening devices incorporating a small grinding wheel are also produced for use with a power drill. Sharpening jigs are also produced into which the drill is mounted and then pushed to and fro over a sheet of aluminium oxide abrasive paper.

It is sometimes none-too-easy to be sure the hole is being made exactly where it is wanted, especially as twist drills lack an actual sharp point. Making a small indentation with a bradawl at the required position ensures the accurate location of the hole. The same dodge can be used for all boring tools where the exact position of the hole is important.

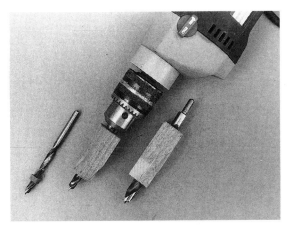

15 *Simple depth control devices*

Lip-and-spur bits

A type of wood boring bit that has evolved from the engineer's twist drill is known as the lip-and-spur. This has a point, which makes location easier, and is ground in such a way that the outer limits of the bit lead the cutting action. Thus the fibres of the grain are first severed before the waste is cut and lifted out. These bits are produced in imperial sizes from $\frac{1}{4}$in to $\frac{3}{4}$in, and metric from 4mm to 20mm, the larger ones having their shanks reduced for gripping in the chuck of the drill. They are usually made from either HSS or chrome vanadium steel, with cheaper ones being produced in carbon steel. They can all be sharpened on a fine grinding wheel. These bits are ideal for boring dowel holes and similar holes of medium diameter and depth.

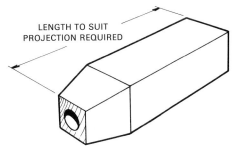

LENGTH TO SUIT PROJECTION REQUIRED

16 *Wooden depth stop sleeve*

Depth control

It is often necessary to bore holes to specific depths, especially when these are in the face of a board and there is little latitude between the depth of hole required and the thickness of the material. Depth control is easy to attain with a drill stand or drill guide, but when working freehand, some form of limiting the depth is needed. The simplest way is to use a rubber collar, often provided with the simpler 'dowelling jig kits' when the bit is also included. These need care in use, as the danger is that they get pushed up the bit as the holes are made. Frequent checking helps to prevent errors. Steel collars are also available. These are locked in place with a grub screw and are made to correspond with the popular dowel diameters of 6mm, 8mm and 10mm ($\frac{1}{4}$in, $\frac{5}{16}$in and $\frac{3}{8}$in). One of the cheapest solutions to this problem is to make a wooden sleeve. A piece of wood of square section

rather greater than the diameter of the bit is bored down its length, then cut to size so that the amount of bit projection equals the depth of hole required.

A depth stop such as the Eclipse, although intended for boring when using a brace, can also be fitted to many bits when used in a power drill. In addition, many drills incorporate a depth stop arrangement, although these are not quite as precise as those mounted on the bit.

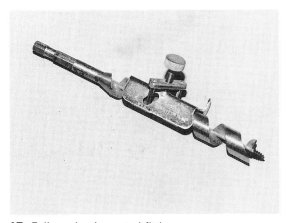

17 *Eclipse depth control fitting*

Flat bits

The flat bit has been specially developed as a simple, economical bit to use with portable power tools, and while it can be used when the drill is in a drill stand, it is particularly suitable for use when the drill is handheld. Its large point gives a lot of stability when used in this mode, and in addition it allows the hole to be made at an angle to the face of the wood. They will bore any kind of wood in any direction of the grain, and cut by a scraping action. It requires relatively little energy for them to function, so quite large holes can be made with drills of modest rating. They are made in both economy and professional quality, and in sizes from ¼in to 1½in along with the metric equivalents. For light and occasional use, it is possible to obtain a flat bit set comprising a single shank with interchangeable bit-ends. The size range of these is limited. The bit slots into the end of the shank and is then held by a single screw. Flat bits are far more efficient in their cutting at relatively high speeds.

Although the accuracy of flat bits is quite good, they should not be used for precision work, and especially for deep holes which are made freehand. This is because controlling the line of the hole is entirely by the hands, and if these move it is possible to produce a hole which follows a curved path. The large point can also be a restrictive feature if blind holes in thin material are required.

18 *Flat bits and extension piece, from Ridgway*

When it is necessary to bore a hole in a particularly restricted space, or where access is difficult, the use of an extension shank for the flat bit might solve the problem. An extension shank will add a further 250mm (10in) to the effective length of the bit, but is not suitable for the smaller sizes of bit. The connecting collar requires a space of 16mm (⅝in) for it to pass through, and shanks can be coupled together for addtional reach.

Through holes

Boring through holes with any kind of bit requires a simple technique to prevent damage as the bit exits on the reverse of the wood. In fact, there is a choice of methods in order to avoid the splintering which would otherwise take place, and these are exactly the same as if the holes were being made using a hand brace. Either scrap wood must be held tight up to the reverse side of the wood and the hole made through the workpiece and part way into the scrap, or the wood must be reversed as soon as the point of the bit emerges and the boring completed from the reverse side. Because of the speed of boring with a power tool, the latter technique is not easy without depth stop provision. If the hole is being made on a drill stand, it is usually sufficient to have the scrap beneath the workpiece – a normal precaution anyway to prevent damage to the base of the stand. Reversing

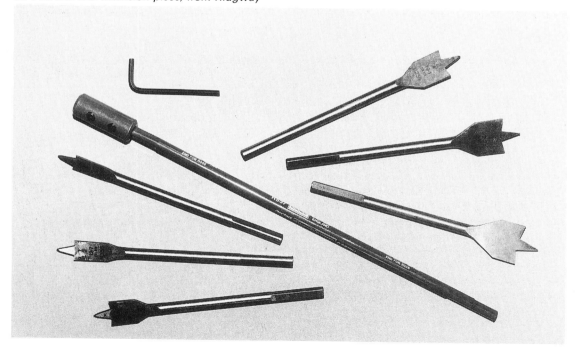

the direction of boring with a flat bit is not recommended, as the bit needs to have its point embedded in wood for proper control to be maintained.

Flat bits can be sharpened on a grindstone. If not too hard, then a file can be used. Although these bits work by a scraping action, there is no need to produce a 'burr' of the type used on a cabinet scraper. A bevel of approximately 75 degrees produces a good cutting action. In addition, it is also possible to grind down the edges of these bits in order to bore a hole of other than stock size diameter. Thus, for example, a standard bit of $1\frac{1}{8}$in can be reduced by grinding to create a bit which will bore a hole of $1\frac{1}{16}$in, or any other size not manufactured. This grinding should not be carried out so as to produce cutting edges.

In addition to being able to reduce their size, you can modify these bits to make a hole other than of cylindrical form. By tapering the edges a tapered hole can be readily formed, or by a combination of tapering and rounding the lower end a tapered hole that also has a rounded base can be produced. Both of these modifications are known to be very successfully used by a woodturner, the first to form a hole to match the tapered end of a decorative candle, and the second to bore out the inside of wooden thimbles. In both cases the slight bevel needed to produce a cutting edge is continued up the edge of the bit, as it is essential for these to cut for the bit to work at all.

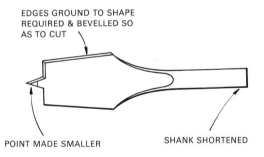

EDGES GROUND TO SHAPE
REQUIRED & BEVELLED SO
AS TO CUT

POINT MADE SMALLER   SHANK SHORTENED

19 *Modified flat bit for special needs*

Bits adapted as above can only be satisfactorily used where the movement of the bit is totally controlled, and because of this the point can be considerably reduced in size. As bits modified for such special requirements will invariably be used only for boring relatively shallow holes, a further aid to accuracy is to reduce the length of the shank by a half or even more. This reduces the slight tendency these bits have to flex, owing to the shank being relatively long compared to its diameter.

Forstner bits

The Forstner pattern bit produces a particularly smooth, flat-bottomed hole. This bit is guided by its sharp-edged rim, its tiny point only being used at the start of the boring. It is not suitable for end grain work, and has limitations on holes deeper than around that equalling the diameter. The full circular rim tends to produce a fine dust, as opposed to chippings. The dust does not escape easily, but becomes trapped between the rim and the side of the hole and overheats because of the friction. This can lead to smoking, or even burning, with certain timbers.

Nevertheless, the Forstner bit does have its uses, especially for shallow holes. The holes are clean and true. It is unaffected by knots or the irregularities of the grain. Overlapping holes can be made, and it is ideal for decorative patterns, scroll-work, and scolloping, and is one of the few bits that will bore veneer. A slow feed rate should be adopted.

The bit is not easy to sharpen, and because of this its use is best restricted to where it scores over alternatives. The inner edge of the rim is sharpened by using a slipstone, and the cutters with a smooth file. They are produced in diameters in imperial and metric sizes from $\frac{5}{16}$in to 2in and 8mm to 50mm, which means that for anything other than very shallow holes in woods up to medium hardness a very powerful drill would be needed for the larger sizes.

Saw-tooth bits

The saw-tooth bit is similar to the Forstner bit, the lower edge of the rim having saw-like teeth rather

20 *Saw tooth bits*

than a continuous sharp edge. This pattern of bit has a very efficient cutting action in all directions of the grain and will bore holes deep and shallow, in hard or soft wood, wet or dry. Made in imperial and metric sizes from $\frac{5}{16}$in to $3\frac{1}{2}$in and 8mm to 75mm, but as with the Forstner bit, and indeed other bits as well, the effective maximum diameters that can be produced with a portable power tool are governed by the power rating of the drill, and the hardness of the wood. These bits can be easily sharpened by file.

Double-cutter bits

A third type of bit that is broadly similar to the Forstner and saw-tooth bit is the double-cutter centre bit. A versatile bit particularly suitable for plywood, the hole produced is cleanly cut largely because of the substantial outer radiused scribers that sever the fibres before the cutters lift out the waste. They are produced in sizes up to 2in and 40mm.

Shank sizes

The Forstner, saw-tooth and double-cutter centre bits all have an overall length of 150mm (6in), and a shank diameter of 13mm ($\frac{1}{2}$in) which is 50mm (2in) long. All these bits perform better when used with the drill in a drill stand because of the very limited guidance and control offered by the business end of the bit. They are at their best when used for fairly shallow holes because of limited chip ejection behind the cutting edge. The RPM of the drill and the feed rate should both be moderate, so that the bit is cutting freely rather than by a combination of high speed and forced rate of feed. The feed should be steady so that the cutting is always being maintained. 'Dwelling' causes the bit to rub and thereby dull quickly.

Auger bits

The machine versions of wood augers are produced in a range of patterns. They are characterized by the screw point and the extended fluting up the length of the bit. The spiral fluting allows the chippings to be drawn out of the hole as boring proceeds, thus eliminating the problem of 'choking', which some bits suffer from when deep holes are being made. The outer surface of the twist on any bit with spiral fluting is known as the land, and on these augers this is quite wide. The land offers support and guidance to the bit, especially on deep holes. This ensures straight boring,

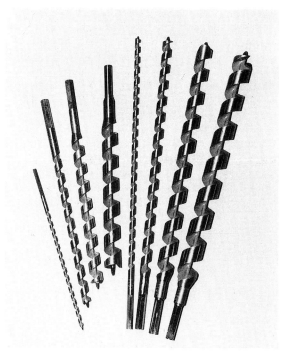

21 *Long and medium length twist bits, from Ridgway*
22 *Short series twist bits*

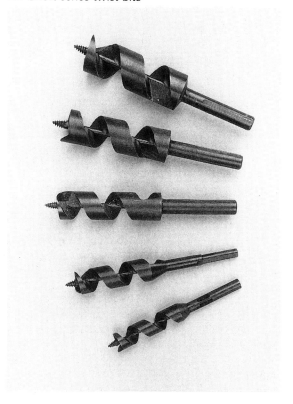

but when used freehand it is essential to align the bit and drill correctly at the start of the operation, as correction during boring is impossible.

Auger bits are made in various 'series', usually according to the overall length of the bit, and are normally classed as short, medium and long. Actual lengths range from 125mm (5in), up to 450mm (18in), although not all diameters are produced in all lengths. In addition, the length of the spiral relative to the overall length can vary even for the same diameter of bit. Most patterns of this bit are made with one scriber and a single cutter, while some versions are produced with two scribers and two cutters. Although the single cutter/scriber auger cuts cleanly and well, those with double cutters and scribers are a little more efficient because of their better balanced cutting action. This is more important on end grain work because of the rather greater resistance the wood offers when being worked in this direction.

Some of these augers are manufactured with round shanks, others with hexagonal. The latter are also produced so that the hexagonal shank incorporates a recessed cylindrical portion part way along the shank. These are combination augers. The shank is designed so that it can be used in either a three-jaw drill chuck or the two jaws of a hand brace.

The screw point of an auger bit is very efficient and when the drill is handheld this feature is very effective at drawing the bit into the wood. In a drill stand, though, the screw point can be almost a disadvantage and cause a blind hole to overrun. Where a screw point bit is regularly used in a drill stand, filing the point to reduce its effect can be an improvement.

Concealed hinge bits

One of the many advances in cabinet hardware over the years relates to the type of hinge generally known as the 'concealed' pattern. Not only are these completely out of sight within the cabinet, they are also adjustable once the door is mounted, which allows for accurate alignment of the door to the carcase. All are adjustable in two directions; some are available that can be adjusted in all three planes. They are made in many variations and finishes, spring-loaded and plain, and in two sizes of 35mm and 26mm. Although used extensively in man-made boards, they can also be adopted for solid wood.

The body of the concealed hinge is a shallow cylinder; therefore, it is necessary to make a hole in the inner surface of the door into which it must fit exactly. Although the hinge is also secured by two or more screws, much of the support the hinge offers the door

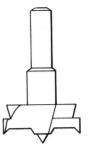

23 *Bit for concealed hinges*

relies on the snug fit of the cylinder in the hole. Accurately prepared holes are therefore essential.

Special hinge boring bits are produced to enable concealed hinges to be mounted, and are made in the two sizes that correspond to the hinges. They have only a tiny point and produce an accurate hole with a flat bottom, essential because of the depth required in relation to the usual thickness of the man-made material often used. Both sizes are made in HSS and TCT. The latter are preferable for use in chipboard and high volume work, and particularly if the board is melamine-faced. These bits should only be used in a drill stand, or at least a drill guide.

Countersink bits

Most screws used for wood-to-wood fixing are of the countersink head type, and therefore the hole made for the screw needs 'countersinking' in preparation for the head. The most popular type of countersink bit is the rose pattern, with multiple cutting edges and an

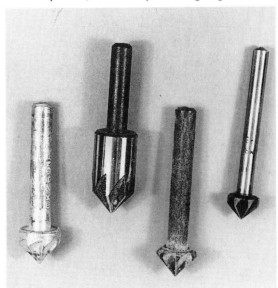

24 *Various patterns of countersink bits*

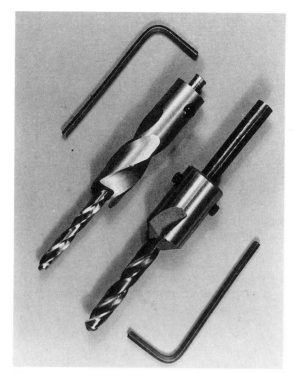

25 *Adjustable counterbore and countersink, from Clico*

Counterbore bits

When the head of the screw is recessed into a hole in the wood instead of being flush with the surface, the technique is known as counterboring. Devices similar to some countersinks are available for counterboring. They are adjustable and shankless, and are fixed onto the twist drills (of specific diameters) with grub screws. The counterbores are available with diameters from 10mm to 30mm ($\frac{3}{8}$in to $1\frac{1}{4}$in). A smaller range of one-piece drill and counterbores are also produced and, again, there is also an economy pattern suitable for occasional use.

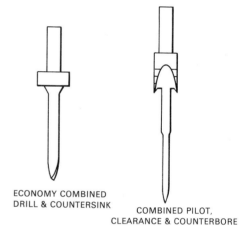

ECONOMY COMBINED
DRILL & COUNTERSINK

COMBINED PILOT,
CLEARANCE & COUNTERBORE

26 *Combination bits for preparing screw holes*

included angle of 90 degrees. Popular sizes of the fixed type of countersinks, in imperial and metric, are $\frac{3}{8}$in, $\frac{1}{2}$in and $\frac{5}{8}$in, and 9mm, 14mm and 20mm. For most woodworking, a larger size is preferred, as this can be used for a range of sizes from small up to its own diameter.

An alternative to the above is the adjustable shankless type. These are made to suit specific sizes of engineer's pattern twist drills, and are secured to the drill by safety head grub screws tightened by an Allen key. This type of countersunk bit allows for the hole and countersinking to be made at the same time.

A variation of the adjustable shankless type is a pattern produced where the drill and countersink are in one piece. Because of the purpose of these bits, they are produced to suit the screws and typical sizes are for number 8, 10 and 12 screws. An economy version of this combined type is also produced in a similar range of sizes. The simple design cuts by a scraping action.

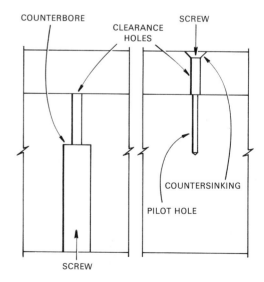

27 *Counterbored and countersunk holes for screws*

If a counterbored hole is being prepared without the use of a special cutter designed for the purpose, then the larger hole must be made before the smaller one. For all practical purposes, it is impossible to make a small hole into a larger one simply by following through with a bit of larger diameter, and this should be remembered at all times, not just when counterboring is required. The exception to this is when using twist drills. These do not present any problems when enlarging is needed.

Plug cutters

Screws are counterbored for various reasons, and often there is no need to conceal the hole after driving in the screw. However, counterboring is also adopted on visible surfaces and where it is better for the screw to be hidden. This means filling the hole with a plug of wood to match that through which the screw is driven. Plug cutters are produced for this purpose, but for a satisfactory fit of the plug within the hole, it is essential that they are correctly matched to the bit used for the counterboring. For kitchen units, bedroom built-in furniture and similar simpler types of constructions, screws which are counterbored are often concealed by a plastic cover cap. Because these rely on a tight, press-in fit, it is wise to trial-bore a piece of scrap to check the fit of the cap within the hole.

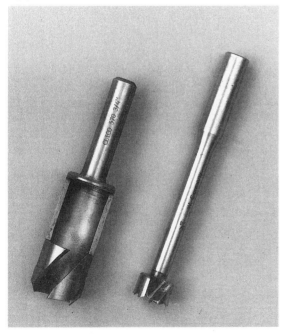

28 *Matched plug cutter and bit, from Clico*

Expansive bits

With such a vast assortment of bits available to suit every boring requirement and size of hole that might be needed, even the professional user would find it almost impossible to have available in his workshop a full range of all types of bit produced. Often holes of a particular size are wanted at only very infrequent intervals. This is specially true of holes of the larger diameters. To buy a saw-tooth bit of large diameter for the sake of maybe a couple of holes can be costly, so for these circumstances the expansive bit offers a worthwhile alternative. A total of four of these bits are produced by Ridgways, two of medium duty and two of a heavier pattern. The two medium-duty bits have cutting capacities from 13mm–45mm ($\frac{1}{2}$in–1$\frac{3}{4}$in) and 22mm–76mm ($\frac{7}{8}$in–3in), and the heavy duty ones from 22mm–50mm ($\frac{7}{8}$in–2in) and 35mm–79mm (1$\frac{3}{8}$in–3$\frac{1}{8}$in). The adjustable cutters are of chrome steel. They are graduated, enabling them to be set to cut a hole of a given diameter, and are replaceable. A drill with a motor of 500 watts is needed for these bits, and even then boring very large holes in fairly tough wood might be beyond all but the heaviest of industrial power drills.

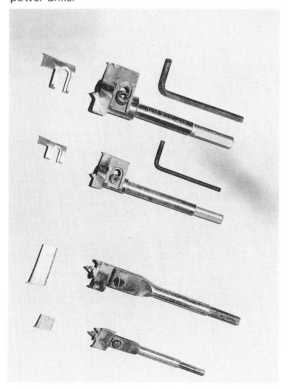

29 *Medium and heavy duty expansion bits, from Ridgway*

Hole saws

Yet another alternative method of cutting large holes is by using a hole saw, although these will only function where a through hole is required. Various patterns and qualities are produced, with the better ones able to tackle various metals as well as wood and plastics. These are of HSS and have welded edges to the cylindrical saw, and are mounted onto an arbor that incorporates a pilot drill. The drills are replaceable, and are held in place by a grub screw in either a 'dimple', or 'flat' on the drill. The single arbor will accommodate all the blades from the range.

Cheaper ones are of carbon steel, and are intended mainly for wood. For limited use, hole saw sets offer an economical means of forming large through holes. These sets allow for up to seven blades of different sizes to be mounted onto the one body, these also having a renewable pilot bit.

As well as for use in forming holes, hole saws can also be of use when it is the disc removed from the hole which is needed, providing that the centre hole is acceptable. Size range across the various patterns of hole saws is vast, from around 14mm ($\frac{9}{16}$in) up to 152mm (6in). Cutting depths of different types can vary up to around 38mm ($1\frac{1}{2}$in). Hole saws should only be used in a stand, and smallish pieces of wood should be cramped, held in a drill vice, or otherwise secured to the base or table of the stand. This is because there is a risk of the teeth of the cutting edge biting into the wood, and causing the wood to rotate. This is especially so the larger and deeper the hole, and a major precaution is to follow the relatively slow recommended speeds of the manufacturers. Often the waste disc once cut is retained within the saw, and has to be manually freed once the drill has stopped.

Hole rasps

Occasionally a hole is needed that is not truly circular, or the edges need to be bevelled or adjusted in order that a component may be fitted within the hole. A hole rasp is useful here. These are around 8mm ($\frac{5}{16}$in) diameter, and while they will actually bore a hole to this diameter, their main characteristics are the rasp-like teeth in the centre part of their length. They are effectively a rotary file, and are used in a freehand manner by the movement of the drill. They are a little rough and erratic in their cutting, and so are only suitable where precision and appearance are not too important.

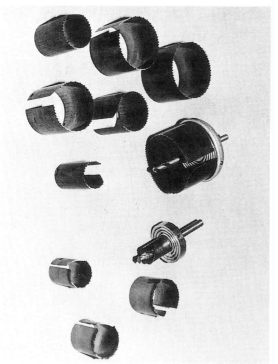

30 *Hole saws with assorted blades*

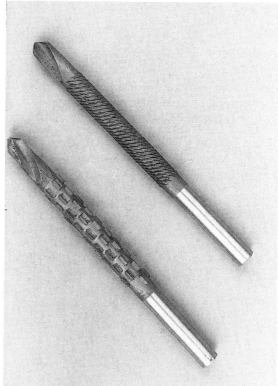

31 *Coarse and fine hole rasps*

Masonry drills

Many items of woodwork have to be secured to walls and other similar structures on completion, normally by some kind of 'plug'. In all cases where a form of anchorage is required into brick, stone, or like material, a hole needs to be made so as to house the actual fixing. Holes into these types of hard and very abrasive materials are made with TCT masonry drills. Various lengths and diameters are produced, although for the woodworker, the shorter lengths and smaller diameters will meet normal needs. The drills are produced in imperial and metric diameters, and also to suit the gauges of wood screws most commonly used for fixing purposes – numbers 8, 10, 12 and 14. Drills which are matched to screw gauges are, of course, larger than the shank of the screw, as they are designed to be used in conjunction with proprietary wall plugs, which are usually plastic. While masonry drills will produce accurate holes in dense materials, there is often the danger of an over-sized hole being made on softer wall materials such as the mortar, or plaster covering.

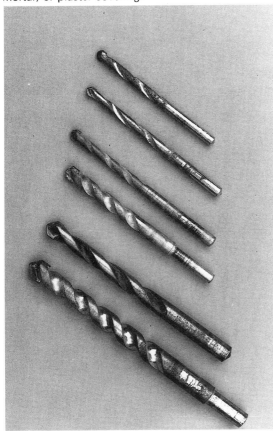

32 *TCT masonry bits*

Masonry drills are produced with TCT of slightly varying form, designed according to the type of power tool they are to be used in. Rotary drills are intended for power tools without hammer action, while impact rotary drills will function in power tools with or without hammer action. Some power drills are manufactured with an electro-pneumatic hammer action, developed particularly for masonry drilling but also capable of carrying out normal boring. These tools have a special chuck for holding masonry drills, which in turn have to be the pattern known as SDS. These have grooves machined into the shanks that enable positive engagement to be made within the chuck and, in addition, the cutting edges have a negative angle.

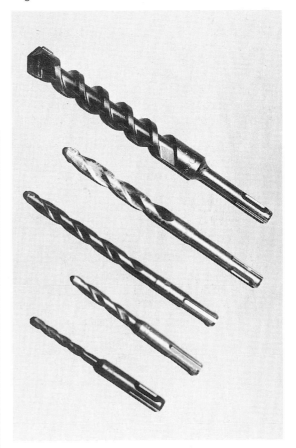

33 *TCT masonry bits for electro-pneumatic drills*

Tungsten carbide is extremely hard and cannot be ground on ordinary grinding wheels. A type of wheel known as 'green grit', or alternatively a silicon carbide grinding wheel, is required for sharpening masonry drills, a characteristic of the operation being the absence of sparks.

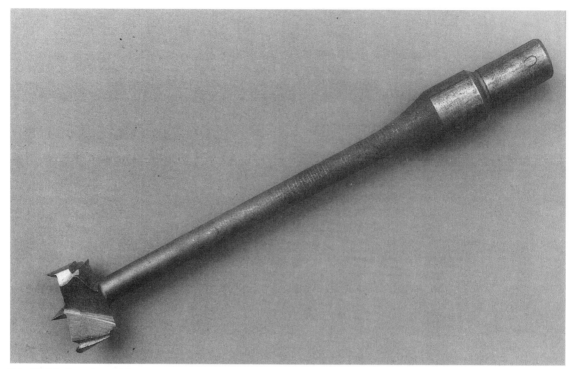

34 *Multi-directional bit*

Multi-directional bits

A bit of Scandinavian origin has been developed that is designed to bore holes that follow a curved path along their length. The bit resembles a double-cutter centre bit, but has cutting edges on its outer, and also its upper, surfaces. It is also possible with this bit to bore part way into the wood, then cut sideways as far as the shank will allow. Thus a hole could be formed which in section would be T-shaped. Trials with a sample of this bit proved to be rather disappointing, and practical needs for holes of this type would, in fact, be extremely rare.

Chuck adaptor

There are occasions when it would be useful to be able to use a power tool bit in a hand brace, but the chuck of a hand brace is designed for the square taper tangs of traditional bits, not those of cylindrical form. There is an adaptor produced that enables this to be done. It is a normal three-jaw chuck with a square taper arbor, which provides a positive hold in a hand brace.

4 ‖ Drill Stands and Drill Guides

By the very nature of its portability the power drill is likely to be used handheld for the bulk of the work it is expected to carry out. When used freehand, though, control of the boring is very much dependent upon the hand and eye of the user, and thus the degree of accuracy with which the tool can be used becomes a matter of skill and judgement. While it is not difficult to control the depth of the boring fairly accurately by the use of depth stops or collars, ensuring the holes are made square to the surface of the wood is far more difficult. In addition, when the drill is used handheld and the holes need to be accurately positioned, all of them will have to be individually marked so as to give their exact location.

36 *AEG drill stand*

By using the drill in a drill stand, many of the problems of accuracy are eliminated. The holes are automatically made square to the surface of the wood, and the stands have a built-in depth-stop provision that is more precise and more reliable than most types which fit on to the bit. By providing a fence to the table of the stand, or by making a simple jig, better positional accuracy is gained and the location of the holes often only need to be partially marked out. Much of the waste from mortises can be easily bored out using a drill stand and a fence. For repetition work, the use of a drill stand is quicker and less tiring than using the drill handheld. The drill stand is particularly useful when holes need to be bored at a specific angle into the face of the wood. While only a small number of drill stands have built-in provision for angle boring, simple jigs can be made for boring the wood at angles

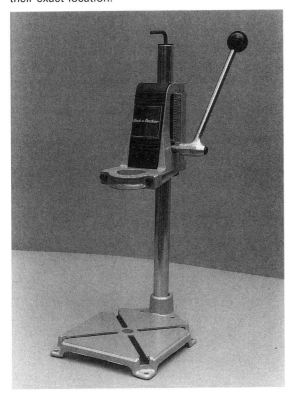

35 *Black and Decker drill stand*

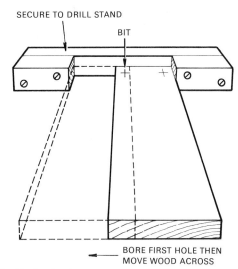

SECURE TO DRILL STAND

BIT

BORE FIRST HOLE THEN
MOVE WOOD ACROSS

37 *Jig for boring two holes at predetermined positions*

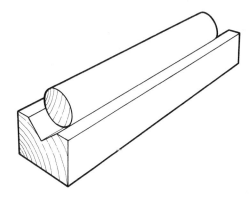

39 *Cradle for holding cylindrical work when boring on drill stand*

other than right angles. Some bits, especially the larger ones such as expansion bits and saw-tooth hole cutters, perform far better in a drill stand where the rate of feed can be more easily controlled. In addition, most stands will accept a router, providing this has a detachable base, and a collar size and pattern compatible with the mounting bracket on the head of the stand. Certain heavy-duty stands are designed so that square chisel mortising can be carried out, for which attachments are specially produced. Some, but by no means all, of the advantages of the drill stand also apply to drill guides. Indeed drill guides can often be used when a drill stand cannot.

Drill stands do have their limitations. The pillar that

supports the drill imposes a restriction on the positioning of the wood beneath the drill, and therefore the location of the holes. Large workpieces can only be held on the fairly small table of the stand if extra support is provided, and boring holes into the ends of the wood, other than quite short pieces, is virtually impossible.

Work-centre mounted drill stand

There is, though, at least one 'work-centre' available that has a hole in its cast aluminium table intended to accommodate the column of a drill stand, so that the table of the work centre is effectively substituted for the base of the stand. The governing factor is the size of the hole, which must match the diameter of the column. The Meritcraft work centre has a 25mm (1in) hole, which corresponds to the column of the Black

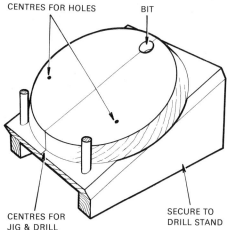

CENTRES FOR HOLES   BIT

CENTRES FOR
JIG & DRILL

SECURE TO
DRILL STAND

38 *Jig for boring holes for legs on underside of stool top*

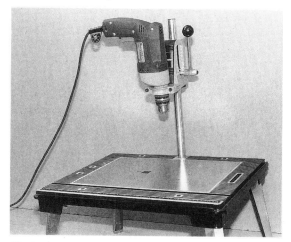

40 *Black and Decker stand mounted into Meritcraft work centre*

and Decker drill stand. The advantages of using a drill stand in conjunction with a work-centre are that the improved size of the table offers far better support to large workpieces, and it offers bench-type drilling facilities if the work-centre is taken 'on site'.

Capacities of stands

Most drill stands are designed to accept drills with 43mm collars, although some heavy industrial models are intended to accommodate corresponding heavy-duty drills with 57mm collars. The throat capacity of a stand is the distance from the centre of the mounting clamp to the edge of the column. As the centre of the clamp corresponds to the axis of the drill and the bit, this throat size gives the maximum distance at which boring can take place from the edge of the wood. This distance varies from stand to stand from around 115mm to 230mm (4½in to 9in), although the popular range of stands have capacities up to around 150mm (6in).

The stroke of the stand relates to the maximum vertical movement of the head, and therefore the drill,

as the lever is brought down. This is approximately in proportion to the overall size of the stand and also the throat capacity, and ranges from around 63mm to 100mm (2½in to 4in). This is adequate for most purposes, and does not impose the same limitations on use as does the throat capacity. All drill stands have the facility for positioning the head at different heights on the column and locking it in place. This does not change the stroke length. The arrangement is to allow for different sizes of workpiece to be positioned on the table.

Movement of head

As well as being able to raise and lower the head, it is also possible to rotate it, with most stands, through 360 degrees. This is done either by rotating the head itself or, with some models, by rotating the head and the column, the column being held by a clamping arrangement in the table of the stand. Some stands have columns that are hexagonal in section and, because of the alignment of the head onto the column, rotational movement can only be gained in steps of 60 degrees.

The advantage of rotating the head is to swing the drill clear of the base. Depending on how the stand is mounted on the top of the bench, this could mean

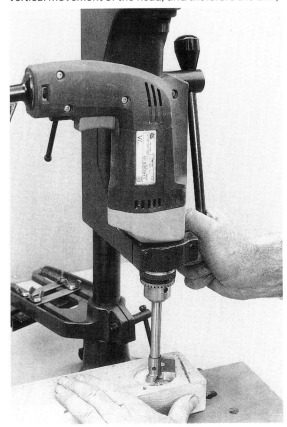

41 *Ridgway heavy-duty universal stand*

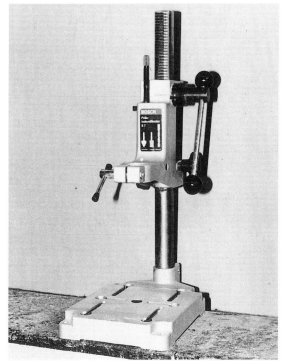

42 *The Bosch drill stand*

unrestricted space between the drill and the floor. Thus large items, or those which require boring after assembly, can be tackled with the drill secured in the stand, an example of this being boring out the edges of doors for mortise locks.

The vertical movement of the head is obtained in different ways on different stands, but in all cases a spring, which is usually a coil type encircling the column, raises the head after lowering, and retains it in the upper position. On some models lowering the head is achieved by a lever-linkage system when the main handle is depressed, while on others it is by rack-and-pinion. Lowering the handle rotates the pinion, with the rack part of the mechanism usually being machined on to the column, although the rack can be a separate component within the head.

Refinements on some models include a graduated scale for use in conjunction with the depth stop, and provision for housing the chuck key of the drill or Allen key needed for making adjustments to the stand itself. All stands need to be secured to a bench in order to be used fully, and this is done by bolting or screwing through the base.

Base of stands

In addition to the holes in the base for holding down purposes, there are either slots or holes or usually a combination of both machined into the base. They enable items to be positively secured to the base to add to the methods of working. A common piece of equipment used on a drill stand is a machine vice. This is really an engineer's item of equipment used principally when pieces of metal are being drilled. Most woodworkers do need to drill metal from time to time, for instance when making simple brackets or adapting hardware, and good machine vices have both vertical and horizontal vee-grooves in the jaws so that bar material can be properly gripped while being drilled.

43 *AEG drill vice*

Drilling metal

It is a dangerous and bad practice to hold metal in the hand when drilling, unless the material is of a good length. The drill can snatch the metal, wrench it out of the hand, and rotate it while locked on the drill. This is particularly likely to happen at the point of break-through on the underside surface. It can result in the metal striking your hand as it continues to rotate, or in the drill breaking and the metal being flung across the workshop. If a machine vice is not available, or the size or shape of the metal makes it unsuitable for gripping in a vice, then the material should be gripped in a mole-type wrench.

44 *Drilling metal using drill vice*

Procedure for use

The machine vice can, of course, be used for holding wood during boring, and is particularly useful for small pieces. Generally, wood is not likely to grip the bit in the same way that metal does, although there can be a considerable leverage effect when boring fairly large holes. This is simply because of the resistance of the wood to being bored and therefore, when large holes are being made in relatively small pieces of wood, it should be restrained in a jig or similar device. Hole saws used in thickish material tend to grip, because of the limited amount of waste removed as cutting advances.

At all times a piece of wood packing should be used beneath the workpiece – essential for through holes but desirable with blind holes as well. All wood boring bits would be severely damaged if they penetrated the workpiece and struck the base of the stand, and

45 *Always use scrap wood beneath workpiece*

engineer's morse drills would cause damage to the stand. It is a wise precaution to set the depth stop to prevent this possibility, even when scrap wood is being used beneath the workpiece.

The slots in the base can also be used for securing home-made fences. These are better if they incorporate their own base piece through which bolts can pass for fixing. When set in place relative to the axis of the bit, holes are automatically bored at a predetermined distance from the edge of the workpiece, useful when preparing holes for concealed-type hinges that need accurate positioning in relation to the edge of the wood. Such a fence provides guidance to the wood when a series of adjoining holes can be made as a simple means of boring out the waste for mortises. Where the drill stand is being used in conjunction with a router, such a fence is essential for much of the work that can be carried out in this way. The fence should be made considerably larger than the base of the stand so as to support the wood adequately. A small number of manufacturers do provide routing/milling benches for their stands that are available as attachments, one such manufacturer being Bosch.

Mortising on the drill stand

For square chisel mortising the stand has to be specially designed for the purpose and of a particularly robust construction. One such stand is the Record Universal Woodworking Stand. The chisel is secured in a special holder with the holder itself being held on the underside of the clamp that grips the drill. Other essential parts of the mortising facility on this stand are the two guides that control the position of the wood beneath the chisel, and the fork bracket on which the guides are mounted. The maximum size of chisel that can be accommodated in this stand is 13mm ($\frac{1}{2}$in). The wood has to be fed manually along as successive cuts are made.

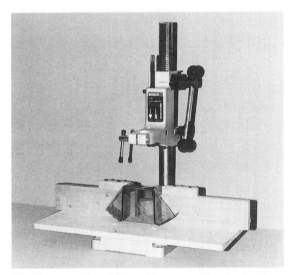

46 *Home-made table and fence on Bosch stand*

47 *Mortising on the Ridgway stand*

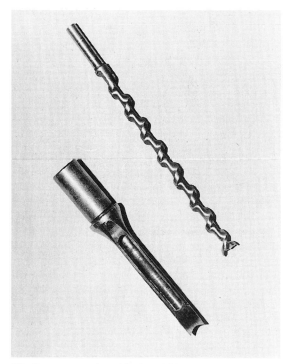

48 *Square mortise chisel and bit*

Square mortising chisels require a special matching bit. These do not have points on the ends and the shank is gripped in the chuck of the drill. When setting up, it is essential that a slight clearance be allowed between the wings at the lower end of the bit, and the bottom edge of the chisel. This needs to be around 1.5mm ($\frac{1}{16}$in), and a simple way of achieving this is as follows. First, secure the chisel into the holder so that there is the required clearance at the top of the chisel between its shoulder and the lower surface of the holder. This gap can be established by placing a coin of suitable thickness between the two surfaces. Next, the bit is inserted into the chuck so that it is tight on the lower edge of the chisel. By slackening the screw that holds the chisel and raising it so that the original gap between the chisel and the holder is closed, the clearance will have been transferred to where it is required at the lower end.

Correct setting of the chisel and bit, together with very firm fixing, are essential. If the bit is incorrectly set, or the chuck is not properly tightened and thus allows slippage, the side lips of the bit will rub on the cutting edges of the chisel. This will result in burning and spoiling of the bit and even possible breakage of the chisel.

The chisel can be locked into the holder in any position through 360 degrees, or effectively in 90 degrees steps as it is square shaped. It is essential to have it secured so that two of its sides are exactly parallel to the edges of the wood, although in practice this is better checked by using a try-square between the guides and the chisel. It is also necessary to ensure that the 'windows' in two of the sides of the chisel face the ends of the mortise, so that the chippings can be more effectively ejected. Clearance is also required between the guides and the workpiece. This must be just sufficient to allow the movement of the wood beneath the chisel, and yet not allow any free play. Free play would result in a mortise being formed that was irregular, and probably oversize as well. To set the guides, a piece of newspaper should be wrapped once around the workpiece, and the guides set to the combined thickness of wood and paper. With the paper removed, the clearance should be just correct. If there is any slack, repeat with thinner paper, or one thickness only.

Mortise chisels and bits are available in metric and imperial sizes, and for small machines such as the Record the sizes which can be used are: 6mm, 8mm, 10mm and 13mm, along with $\frac{1}{4}$in, $\frac{5}{16}$in, $\frac{3}{8}$in, $\frac{7}{16}$in and $\frac{1}{2}$in. As these chisels are made in different 'series' according to the type of machine they are to be used in, it is essential to ensure that the chisels are correctly matched to the machine. Slot-type mortise cutters are not really suitable for use in a drill stand.

It is, of course, possible to form mortises that are wider than the $\frac{1}{2}$in maximum size of chisel suitable for use in the stand. This is achieved by taking a second cut alongside the first. The best procedure is to use a chisel that is half the size of the required mortise, so that both of the cuts are of equal size and therefore waste is being removed across the full width of the cutting edges. Thus for a $\frac{3}{4}$in wide mortise, a $\frac{3}{8}$in chisel should be used, rather than a $\frac{1}{2}$in which would only cut at half its width on the second pass. Any tool cutting wood tends to take the line of least resistance, and therefore the $\frac{1}{2}$in chisel cutting at less than its full width would tend to move slightly into the space already cut, either by straining the chisel, or by forcing the wood a little out of alignment.

For the same reason, the first two cuts made in any mortise, regardless of the width, should be at the ends. If the sequence is reversed, a partial cut made at the ends can cause the wood to move sideways, resulting in the mortise having sloping ends.

A drill stand used for mortising cannot be expected to make the same heavy cuts that are possible with a static mortising machine. The mortise is best formed by taking a series of shallow cuts, especially when working in hardwoods. Through mortises are always made by working from both surfaces, and the depth of stopped mortises is controlled by the depth stop.

A special tool resembling a countersink is required to sharpen the chisel. This tool ensures that all four cutting edges are evenly sharpened and maintained to the correct shape and bevel. These sharpening tools are produced in two patterns. One type has a fixed pilot that exactly fits the bore of the chisel, and thus is only suitable for the chisel of the size that matches the sharpening tool. The second type is produced as sharpening sets comprising a sharpening tool with interchangeable pilots held in place with a grub screw. The smaller of the two sets available covers all the five sizes of bits that can be accommodated in the Record drill stand. To use, the chisel is held in the vice, and the sharpening tool fitted into the chuck of a power drill. Several revolutions of the drill will renew the cutting edges. The inside corners require completing with a fine file. The outside of the chisel must not be filed or honed in any way during sharpening.

To sharpen the spurs of the bit, a small square or flat file is required, and filing must be confined to the inner surfaces only. For sharpening the cutters, a half-round, smooth file is used, with the sharpening taking place from below. At all times, filing should be kept to a minimum consistent with maintaining efficient cutting edges.

On the stand being described, the forks that carry the guides are simply swung out of the way when normal boring is required. As the head as well as the forks can rotate to any position on the column, mortising can take place clear of the base of the stand and thus larger workpieces can be tackled.

49 *Mitre attachment on Ridgway stand*

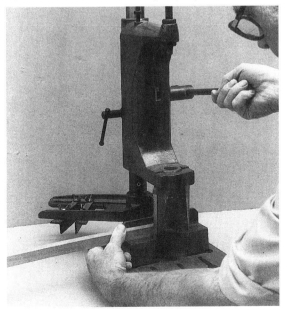

Mitre cutting

The Record Universal Woodworking Stand has as additional facility that is rare and possibly unique to this type of equipment, although it is not power-tool operated. The drill stand will accommodate a mitre cutting attachment consisting of a twin-bladed cutter, which is gripped in the underside of the clamp, and a stepped table, which bolts on to the base of the stand. The mitre cutter will cut wood up to approximately 38mm (1½in) square, providing reasonably thin slices are removed at each stroke.

Electronic machining centre

A very sophisticated type of drill stand is the Wolfcraft model 5005, known as the electronic machining centre. Fitted to the base of the stand is a table measuring 460mm by 240mm (18in x 9½in), which can be moved both laterally and forwards/backwards by handwheels located at the front and right-hand

50 *The Wolfcraft 5005 machining centre*

side. The head will tilt fully to the left or right, and the clamp can rotate through 360 degrees. Thus with a power tool mounted in the clamp, the downwards movement can be made at an angle in line with the axis of the tool, or vertically with the tool mounted at

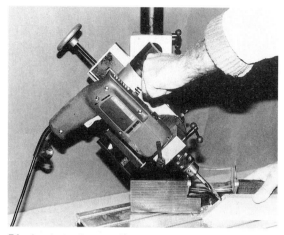

51 *Angle boring on the 5005*

an angle to the downwards path. These angular combinations are especially useful for a router.

The electronic features of this stand are the instant liquid crystal display (LCD) readouts shown in the dials adjoining the handwheels, and a similar dial fitted to the head. Thus the movement of the table, and therefore any workpiece, is accurately controlled by the wheels, with the movement being displayed by the dials. Batteries are required to activate the electronics, both dials fitted to the table showing the movement to 0.1mm. The dial on the head, which shows the movement of this part of the stand, records this to 0.5mm. All the dials can be brought back to zero at any point on their travel by simply pressing the reset button.

The top of the movable table has a series of T-slots in it, as well as one on the front and rear edges. This makes it particularly easy to fit a machine vice, and the two universal clamps supplied as standard equipment with the stand are also anchored by single bolts to one of the T-slots. The clamps are highly flexible in the way in which they can hold workpieces or jigs, and it is by having the workpiece firmly secured to the table that the precision nature of this machining centre can be fully exploited. For instance, if a series of holes with their centres exactly spaced is required in a component, this can be accurately achieved as the handwheels are rotated and the dials read. The amount of travel of the table is very approximately equal to half its size, so the machining centre is at its most advantageous with small size work. Even with larger pieces, the forwards/backwards movement of the table gives precise control to the position of the fence beneath the bit, and therefore very exact locating of the holes relative to the edge of the workpiece is easy to achieve. The fence has a stand, and can be used for boring and routing.

52 *Work-clamping facilities of the Wolfcraft 5005*

As well as the LCD readout on the head by which the downwards movement can be instantly assessed, there is also depth stop provision, so necessary for repetition work. The head will rotate on the column, and can be locked at any height on it. In addition, a handwheel on the top of the head provides fine control as a further refinement to correctly setting the head.

Router work

All drill stands offer the possibility of use for overhead router work. Some have the fence and supplementary base available as attachments, but simple home-made fences can be constructed for others. Further information on using drill stands in this way is given in Chapter 10.

General

What has to be remembered at all times with power tools is that when they are being used in a drill stand or similar alternative means to handheld use, their capacities are not changed. Although it is often more convenient to use power tools in a static manner, thus increasing their flexibility, the power output of their motors remains the same. When power tools are used in the hand, there is considerable sensitivity to their performance, rate of feed, and danger of overload. Some of this sensitivity is lost when the power tool is performing more as a machine than a tool, and the user must be aware at all times of the danger of overloading a power tool in terms of capacity, rate of feed, or its physical and mechanical construction.

The above applies to the drill stands themselves, and indeed all the equipment used in conjunction with power tools. A powerful drill mounted in a small stand could well strain the stand if the combination was repeatedly used for heavy boring work. There is a danger of the drill being forced upwards out of the clamp as the head is lowered on to the work, allied with the possible malfunctioning of the clamp if over-tightened.

Drill guides

The drill guide performs some of the functions of a drill stand, but in a much more simplified way. They are designed as attachments to the drill, and are usually mounted by a clamping arrangement which is tightened onto the collar of the drill. Drill guides are included in the Black and Decker, and the Wolfcraft range of accessories, and both carry out a similar range of drilling operations.

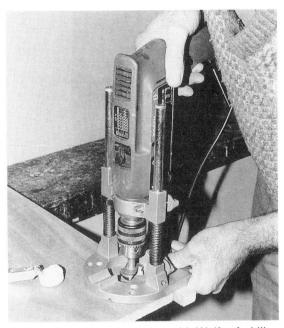

54 *Preparing holes for hinges with Wolfcraft drill guide*

The collar, and therefore the drill, slides up and down the two rods that are attached to the base. Springs on the column raise the drill after the boring, the downwards movement being simply achieved by hand pressure. Adjustable clamps on the rods can be set to limit the movement, and indeed the collar can be locked onto the rods. In its simplest mode of use, the guide and drill are located over the centre of the hole required, and the drill lowered until the stop is reached. This has a particular advantage compared with a drill stand, insofar as the restriction of the column of the latter does not exist, and therefore holes can be made at any position on the face of the work. Both guides allow for angle boring to be carried out, although the means of achieving this differ slightly.

Both guides have V-shaped supports on the upper surface of the base. These are intended for holding round material so that the drill can be lowered and holes made on the centre line of the workpiece. The Wolfcraft guide also allows for the simple reversal of the base on the rods, so that the guide can be used to straddle round work as an alternative means of boring material of this shape. Holes in the base of both models allow for screwing to the underside of a home-made table, so that the drill mounted in the guide can be used as a simple router with a milling-type cutter in the chuck. The table does, of course, require a hole in its centre for the cutter to pass through, and an adjustable fence.

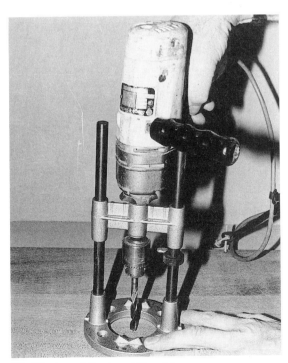

53 *Black and Decker drill guide*

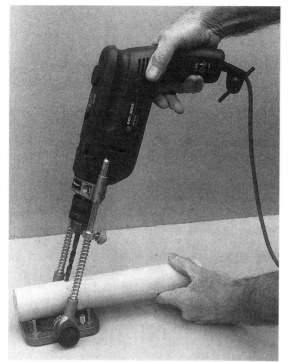

55 *Angle drilling with Black and Decker guide*

A drill guide can be used in a very effective, although limited, way as a dowelling jig. For boring the holes on the face of the work, a batten is cramped to the material to act as a fence for the base of the guide. The batten is secured parallel to the centre line of the holes so that, when the base abuts against it, the bit is directly over the line. The positions for the holes do need to be marked on the wood. The guide ensures that the holes are in a straight line and their depths are easily controlled.

Corresponding dowel holes can be readily made on the edge of the wood, providing they are required in the centre of the thickness. On one of the guides being described, the main rods are set to protrude through the base by around 12mm ($\frac{1}{2}$in). On the other guide, two studs are screwed into the base so as to project by about the same amount. In use, the base is placed on the edge of the workpiece then rotated a small amount until the rods or studs rest against the wood and so prevent further rotational movement. Providing the rods or studs are tight against the opposite sides of the material, the holes will automatically be made in the centre of the edge.

Drill guides used as aids to dowelling operate quite well for box carcase constructions, especially when made out of chipboard or other man-made board. They are not suitable for forming dowelled joints in frame-type constructions where the material is of fairly small section.

56 *Preparing dowel holes using Wolfcraft guide*

57 *Automatic centring of edge holes is achieved with drill guide*

5 ‖ Attachments and Accessories

During the early days of the development of the power drill, along with its acceptance by all who work in wood whether amateur or professional, we also had following very closely behind a variety of attachments that could be added to the drill. In some cases, the attachment effectively converted the drill into another type of tool to perform basic operations such as jigsawing, circular sawing, and sanding. Other pieces of equipment, including the lathe and dovetailing jig, really used the drill as the power source.

To a large extent, the ongoing development of an ever-growing range of portable power tools designed for specific operations has led to the decline of many attachments that had been introduced before single function tools became more widely available. Despite this, the range of equipment that is currently available for use in conjunction with power tools continues to grow, most designed for use with the drill. Equipment developed to be used with routers is already very considerable and continues to expand rapidly. Routers and associated equipment are dealt with in Chapter 10, and specific information on circular and jig sawing, abrasives and sanding will be found in Chapters 5, 6 and 7 respectively.

Only a very small number of manufacturers produce attachments that convert the power drill into a different type of power tool. Inevitably, the capacity and performance of the attachments are largely dependent on the quality of the power drill on which they are mounted, but even with the best of drills, it cannot be expected that an attachment will perform as well as its single function equivalent. Convenience of use, balance, and positions of handles all suffer to some extent on attachments, but nevertheless, they are all capable of carrying out worthwhile work.

Bosch produces attachments that enable sanding, jigsawing, and circular sawing to be carried out, and all these attachments are designed for any drill with a 43mm collar. The sander is of the orbital type, and designed to accept a half sheet of abrasive paper. What has to be remembered with all orbital sanders is that, while they only remove a fairly small amount of

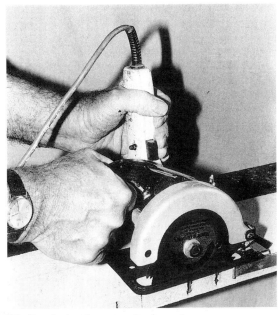

58 *Small circular saw attachment*

wood, they can produce particularly smooth surfaces. Because of this they are often referred to as finishing sanders.

The jigsaw attachment from the Bosch range will cut up to 51mm (2in), providing it is properly matched to a drill of appropriate speed and power. The sole plate will adjust up to 45 degress in both directions enabling bevel cuts to be made. Circular saw attachments inevitably have blades of relatively small diameters, largely because a circular saw requires a lot of power when cutting at its maximum. Even so, this attachment has a generous cutting depth of 42mm (1⅝in) from its blade of 150mm (6in) diameter. A retractable guard is fitted. The sole plate will adjust to provide bevel cutting up to 45 degrees, and in addition the blade projection can be adjusted enabling cuts less than the maximum to be made. A small adjustable fence is also provided, so the attachment has all the essential features of a purpose-made circular saw. For maximum efficiency a circular saw requires a high speed, and therefore a high revving drill is needed for the attachment to operate at it best. All the above Bosch attachments require the chuck to be removed from the drill with the connection then made directly to the threaded spindle in order to transmit the power, with the physical linking being made via the collar of the drill.

Black and Decker also produce similar attachments to the three briefly described above, but they are designed to be used only with Black and Decker drills of the older type, known as claw-grip nose drills. The saw attachment has a blade diameter of 125mm (5in), giving a maximum depth of cut of 30mm ($1\frac{3}{16}$in). These Black and Decker attachments are also designed to have the power link-up directly to the spindle of the drill once the chuck has been removed. Black and Decker drills of the claw-grip pattern have a female thread on the spindle.

60 *Two patterns of rebating attachments*

61 *Wolfcraft rebating attachment in use*

59 *Black and Decker orbital sander attachment*

Rebating attachments

Rebating attachments are in essence small circular saws, although they are not a scaled down version of what has just been described above. With some models, the only point of connection of the attachment to the drill is the saw spindle being gripped directly into the chuck, and while it might appear that the whole attachment would revolve when the drill is switched on and thus control be difficult, this is not the case. The sole plate of the saw has a fixed fence – it is the blade that can be adjusted so that cutting may take place up to around 25mm (1in) from the edge of the wood; operating beyond this limit is impossible.

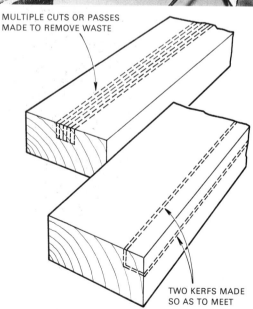

MULTIPLE CUTS OR PASSES MADE TO REMOVE WASTE

TWO KERFS MADE SO AS TO MEET

62 *Forming grooves and rebates with rebating attachment*

Blade diameter on these attachments is approximately 45mm (1¾in), depth of cut adjustable up to around 16mm (⅝in). Thus rebates can be readily formed by cutting from adjacent surfaces with the settings adjusted so that the two kerfs just meet in the inner corner. Grooves can also be formed easily with these attachments. For this type of cut a series of passes is made until the required width is reached.

Rebating attachments are very good for stopped grooves and rebates and, because of the small diameter of the blade, the effective sawing takes place almost completely up to the point where the cut has to stop. This leaves only a small amount to be completed by chisel. These attachments work well in solid wood and all types of man-made boards, although the latter have a marked blunting effect on the teeth. The attachments are particularly useful on chipboard.

Milling cutters

With most of these rebating attachments it is possible to replace the circular saw blade with other types of cutters so that various profiles may be formed in the wood. These cutters are often known as milling cutters, and are shaped so as to produce specific profiles on the wood such as rounding, tonguing, beading, chamfering and similar cuts. It is also possible to obtain grooving cutters of fixed widths as an alternative way of grooving to using the circular saw blade.

Milling cutters usually have nine cutting edges, and are at their best used in a drill of fairly high speed and power. Even so, with certain cuts, and especially in hard woods, it is advisable to form the cut by taking two or even more passes. Speed of working with most of the cutters is relatively slow, and the finish produced on the wood will generally not be as smooth as similar cuts made with a router. Milling cutters are not really suitable for chipboard.

It is possible to use these milling cutters when the drill is in the fixed position, and the wood moved across the cutters. The drill is best held in a drill guide, and a simple table and adjustable fence made so that the drill and its support can be mounted vertically beneath the table. A hole is needed in the table through which the shank of the cutter can protrude, this set-up operating on the principle of the spindle moulder machine, a method of working which is becoming increasingly popular for the router. Some types of rebating attachment are designed so that in addition to handheld working, the whole device can be secured to the edge of a board or table and used in the stationary mode. A guard is provided, along with a special shaped fence allowing curved work involving quite small radii to be tackled.

Milling cutters provide an economical means of forming certain basic cuts, but are only really intended for light and occasional use rather than extended runs. They cannot be regarded as a simplified alternative to similar cuts made by a router, which is capable of a far higher level of performance, accuracy and smoothness of cut.

Flexible drive shafts

Sometimes the physical size of the drill, or its weight or handling characteristics can restrict its use, and what is needed is something smaller and lighter, and more responsive to manipulation. This is where flexible drive shafts can prove advantageous, most of them being intended for light rather than heavy application. Usual lengths for flexible drives are 900mm–1,300mm (36in–51in), with a 6mm (¼in) shank for gripping in the drill's chuck, and either a 6mm or 8mm

63 *Wolfcraft attachment set up for stationary use*

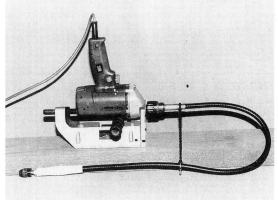

64 *Flexible drive shaft*

($\frac{1}{4}$in or $\frac{5}{16}$in) chuck at the operating end where a handle is fitted. The chucks can be hand-tightened only, this in itself restricting the amount of power that can be transmitted. Whereas most of these shafts have only plain bearings, others have dual ball bearings for longer life and higher operating speed. One model on the market has a threaded end for linking directly into the drill once the chuck has been removed.

Shafts are normally designed for right-hand, or clockwise, rotation only, and they should never be overloaded. Acute bends in the shafts must be avoided. These can result in overheating and rapid wear. The power drill to which it is attached should be held in a drill stand, and not simply trailed behind the shaft when in use. While a shaft can be used for drilling, only small holes can be tackled and, in any case, the shafts are not at their best when axial pressure is required. They are ideal for use with small sanding devices such as bobbins, flap wheels and pads.

Wire brushes

Wire brushes are usually associated with motor cars and metalworking, especially where cleaning and de-rusting are required. They do, though, have a particular use in woodworking, as they can be used to texture a surface. Only certain woods respond to this treatment, usually softwoods. It is essential that the timber has marked spring and summer growth rings, as the wire brush abrades away the softer areas of growth that take place in the early part of the year, leaving the harder summer growth largely unaffected. Work must take place with the grain, or tearing and scoring will result. The flexible shaft provides a useful means of carrying out this texturing. Some experimenting will be needed to get a satisfactory effect, and results are also dependent on the position of the annual rings relative to the surface of the wood.

Lathe attachments

Although a lathe powered by a drill is inevitably small, such a lathe has all the essential features of far bigger machines, and is capable of producing useful components. Some skill and know-how are needed to carry out woodturning in a proficient manner, but a small lathe can be the way to gain practice and acquire these basic skills.

Two dimensions are important relating to the lathe and refer to the size of wood they can accommodate. The distance from the driving centre to the tailstock centre gives the maximum length of wood that can be turned, and is known as the 'distance between centres'. The second dimension is the distance from the bed up to the centres, as this determines the maximum diameter of wood that can be tackled. This is known as the 'centre height', or when this is doubled, as the 'swing' of the lathe. No lathe, least of all a power-drill lathe, can handle a piece of wood whose dimensions equal both the maximum length and swing of the lathe. Space is needed for the tool rest support, and the weight of such a piece would be excessive. Work that is fairly large in diameter but relatively thin is usually mounted by screws on to a faceplate that is supported only on the headstock. Such a piece of wood might be needed for a small bowl or dish, and is known as faceplate turning. With faceplate turning, the grain of the wood is at right angles to the axis of rotation. Wood held between the

65 *Wolfcraft drill stand with wire brush in chuck*

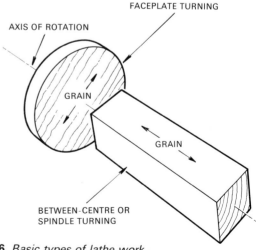

66 *Basic types of lathe work*

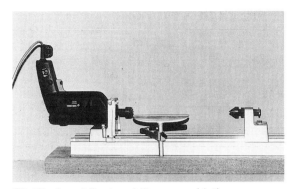

67 *Black and Decker drill-powered lathe*

centres is referred to as 'turning between centres', also as 'spindle turning', and has its grain parallel to the axis of rotation.

Lathes produced by Black and Decker, and Bosch have similar capacities, accepting wood up to a maximum length of around 600mm (24in), with 150mm (6in) as the largest diameter that can be turned. The lathe from the Wolfcraft range does not have an integral bed, the headstock, tailstock and toolrest support being supplied as separate units. These clamp onto the edge of the bench or suitable board, which therefore becomes the bed, and thus in theory will cope with work longer than 600mm (24in). In practice, though, other factors govern what is a reasonable capacity for this type of lathe – the weight of the wood if of fairly large section, and the problems of a long workpiece flexing if too slender. An alternative type of lathe is also produced by Wolfcraft. This has its own headstock bearings so that the side thrust imparted to the headstock during actual turning is taken by this bearing, and not those of the drill.

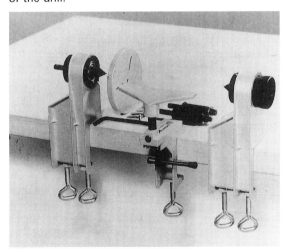

68 *The Wolfcraft lathe components*

Apart from the one just mentioned above, the others are designed for the drill to be mounted in the headstock casting, and the driving centre and faceplate are mounted directly on to the spindle of the drill. It is essential to use a fairly heavy-duty drill for powering a lathe. There must not be any play whatsoever in the spindle as this will result in the wood vibrating as turning is attempted. At least two speeds are needed. Small diameter work needs a high speed, while for large diameter workpieces fewer rpms are required. A speed range from around 1,000rpm to 2,800rpm is ideal.

The tools used for turning are quite different from those used for other branches of woodworking. Indeed, those used for spindle turning are not the same, nor are they sharpened in the same way, as those for faceplate work. The basic tools for spindle work are the roughing gouge, the spindle gouge, the skew chisel, and the parting tool. For faceplate turning, most of the cutting is carried out with fairly small gouges but with deep flutes, and also scrapers of various profiles. Shapes and sizes of all turning tools vary considerably.

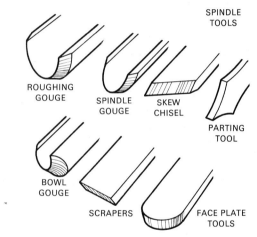

69 *Basic turning tools*

For turning between centres, any material larger than 38mm by 38mm ($1\frac{1}{2}$in × $1\frac{1}{2}$in) should have the corners planed off so as to produce an octagonal section. This reduces the amount of waste that has to be removed on the lathe and lessens the stresses on the equipment. The centre of the wood should be marked at each end with a bradawl. At the end to be mounted in the headstock, one or two saw kerfs might be desirable so that the driving centre, either the two-prong or four-prong type, can gain a positive engagement with the wood. Mount the wood between the centres, tighten the tailstock until the wood is being positively gripped, then slacken off a quarter turn

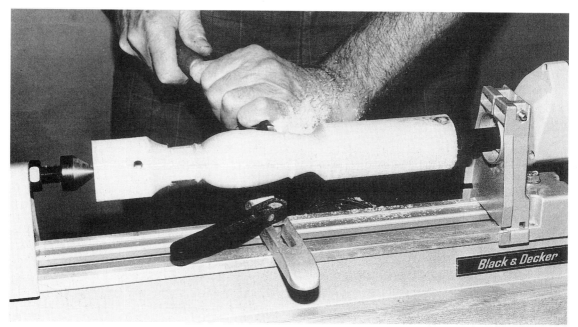

70 *Typical spindle turning*

before locking. Ensure that all nuts and clamps are locked before operating the lathe.

The tool rest should be set at or just below the height of the centres, and about 6mm (¼in) away from the wood. Always rotate the wood by hand before switching on to ensure it is not being fouled. The first stage is to produce a cylinder. This is achieved by the roughing gouge. Cutting must always be carried out as high on the wood as possible. That is, the gouge is inclined upwards, and also points slightly in the direction of working. During turning, the tools must be held positively and yet freely, as constant changes of angle are usually needed as the cutting proceeds. The left hand controls the blade and ensures it is firm against the tool rest. The right hand is on the handle and governs the angle of the tool to the wood, the rotation of the tool and the sideways movement.

The two basic cuts for spindle work are hollows, which are formed with the gouge, and beads, which are formed with the skew chisel. Hollows are started with the gouge well over on its side, and it is rolled over and the handle lowered as the cut is made from each side towards the centre. Cutting must always be with the grain which means on spindle work cutting is from large diameter to small diameter.

For most turning cuts, the bevel of the gouge or skew chisel must rest on the surface of the wood to enable proper control to be exercised. The bevel acts as a fulcrum, and thus enables the depth of cut to be

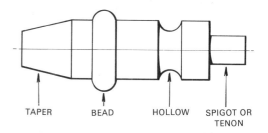

71 *Common cuts in spindle turning*

controlled. This is a golden rule of turning, and is the basis of good technique.

The skew requires practice and skill to use properly, with only the lower half of the cutting edge, known as the heel, used when forming beads, tapers, and smoothing cuts. If the pointed or toe end of the skew contacts the wood when making the above cuts it is almost certain to result in a 'dig-in' to the surface and cause damage.

The parting tool is used to establish the overall length of a turning by cutting in to create a groove and therefore a square cut or shoulder. It is also used to form similar cuts that are a part of a feature, such as a circular peg or tenon. Because the cutting edge operates across the grain of the wood, the parting tool is not the most efficient of tools and cuts made with it can often result in rather ragged surfaces of end grain. If required, these surfaces can be lightly cut with the toe of the skew chisel, with the tool resting on its edge.

If gouges, skew chisels and parting tools are used when resting horizontally on the tool rest they will not cut the wood cleanly, but tear the grain instead. Used this way they simply scrape the wood and yet are not shaped nor prepared as proper turning scrapers are. Scraping tools are not generally used for spindle work.

72 *Face plate work*

Much faceplate work is carried out in two stages: the outer or underside first, after which the wood is reversed and remounted to complete the remaining surfaces. Most bowls are produced this way, starting with a block of wood cut to approximately circular shape. A turned base, as might be required as a part of a reading lamp or stand, could be turned from one side only if the underside was planed flat at the outset. The commonest way of mounting the workpiece on the faceplate is by screwing. Care is always needed with screws to ensure that the tools will not strike them during turning, especially when the wood is reversed and is to be hollowed out as for a bowl. Work that is smaller than the faceplate is first mounted on a packing piece to keep it clear and thus give tools unhindered access to the surfaces near the faceplate.

Gouges for faceplate work are usually sharpened to a fairly steep angle of around 45–60 degrees, and are used straight from the grinding wheel as it is the burr so produced that does the actual cutting. In use, the bevel must again rest on the wood. The gouge is used well over on its side with the angle of the tool to the wood when viewed in plan usually approaching 45 degrees.

Scrapers also cut because of the burr produced when sharpening, and usually cut best when pointing

slightly downwards. Only light cuts should be taken; if the scraper is kept sharp then the waste will be removed as shavings. Dust only indicates the scraper is blunt. Softwoods cannot be scraped. The harder the wood, the better it responds to the scraping technique.

Small pieces of wood such as are needed for door and drawer knobs are most conveniently mounted on a screw chuck which is a common piece of equipment with most lathes. With screw chuck mounted work, the grain can run in any direction, and it is the direction of the grain that determines the method of working and the particular tools used.

73 *Small work held on the screw chuck*

Abrasive paper should not be used until the tools have produced a good surface as abrading will only improve a surface that is already satisfactory. Abrasive paper cannot be expected to make a good surface out of a poor one, nor can it be properly used for shaping the work. When using abrasive paper, remove the tool rest and have the fingers trailing in the direction of rotation.

For even the lightest of turning, the lathe must be properly secured to a bench of reasonable strength and weight. The accepted criteria for a generally satisfactory working height is to have the lathe centres level with the elbow when in a normal standing position.

Inevitably, only the very briefest guidance has been given to using the lathe. Woodturning is a vast topic and craft in its own right, and anyone particularly interested in developing this aspect of working wood should read and study any of the many books on the subject, especially one which deals with 'techniques' rather than 'projects'.

Sanding attachment for lathes

Large, self-powered lathes often have disc-sanding attachments that can be added to them, and a similar facility can be easily made for a drill-powered lathe. A ply disc of around 10mm (⅜in) thick should be screwed to the faceplate, and turned to the maximum diameter possible. To this is added the abrasive paper. A small table also needs to be made and this is anchored to the bed of the lathe. The length of the table should preferably be a little more than the diameter of the disc, with the height of the upper surface being made level with the lathe centre height.

75 *The Wolfcraft multi-directional drill clamp*

74 *Home-made sanding attachment on the Black and Decker lathe*

Drill stands

Drill-powered lathes are often designed so that the headstock can be used independently as a bench stand for the drill, and this can be used in a similar way to the above as the basis for a disc sander. This has the advantage that a larger disc of ply can be used on which to mount the paper, a diameter of 180mm (7in) being about the maximum for safe working. It will still be necessary to screw the ply disc to the lathe faceplate and the sanding table will have to be secured to the bench.

Dowelling jigs

Dowelling is a very old way of jointing wood, and possibly more widely used nowadays than at any time. Originally when dowelling, all the holes had to be separately marked and hand-bored by brace and bit. Its present popularity is largely due to the avail-

ability of a number of dowelling jigs and, of course, the power drill for actually making the holes. Modern glues also make the dowel joint more successful than hitherto, and dowelling is particularly suitable for use in chipboard where many other joints are not. Indeed, although there is a wide assortment of jigs on the market, many of the simpler ones have been designed with chipboard in mind. They can, of course, be used for solid wood, but are primarily intended for boards rather than framing members.

The Spiralux Dowel Mate caters for holes 6mm, 8mm and 10mm, the three pairs of bit-guide holes in the jig being prefixed in their distance from the datum surfaces, as the jig is designed for preparing joints in material of 12mm to 14mm (½in to 17/32in) 15mm to 17mm (9/16in to ⅝in) and 18mm to 20mm (11/16in to ¾in). In use, the two components are secured together by the jig itself aided by a couple of G-cramps, with spacers being provided to keep the material in its correct relative position. In common with all dowelling jigs, the holes in the die-cast jig that guide the bit have hardened steel liners in them to resist wear.

The jig is only fully operational when the holes required are either on the edge of the wood, or

76 *The Spiralux dowelling jig*

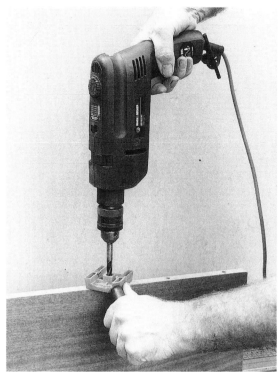

77 *The Wolfcraft single-hole dowelling jig*

immediately adjacent to it. For holes positioned away from the edges, such as might be needed for shelves, the marker pins provided with the jig are used. These are temporarily inserted into the first set of holes made in the end of the component, aligned with the second member, and the positions of the remaining holes marked when pressure is applied.

The Spiralux Mini Dowelling Kit is very simple. It caters for a hole size of 8mm as would be needed in material around 16mm thick. A pair of two-hole boring guides are included with the kit, and some handholding of these is initially required. Two aluminium pins are also provided to locate the guides in the first set of holes while the remainder are made.

Wolfcraft produces several dowelling jigs. The Hobbymate is offered in three different sizes – 6mm, 8mm and 10mm, these being separate jigs. Each can be used on any thickness of wood up to around 30mm (1$\frac{3}{16}$in). In use, the jig is rotated on the edge of the wood so that the two lugs are tight against opposite sides of the board. The holes are then automatically bored in the centre. With the dowels temporarily inserted into these holes and the components aligned

and cramped together, the slot in the fence of the jig straddles the dowels and thus the position for each hole of the second set is located.

The Dowel Master follows the same principles of operation as the Hobbymate, but is designed to cope with the three commonest sizes of dowels, these being 6mm, 8mm and 10mm. Both these jigs from Wolfcraft are of plastic. The Wolfcraft Universal Dowelling Jig is a more substantial tool of zinc alloy, and again provides for making holes of the three popular sizes. Two corresponding holes can be made

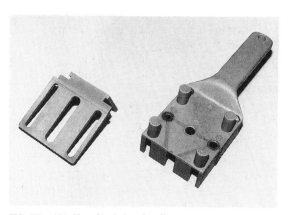

78 *The Wolfcraft triple-size jig*

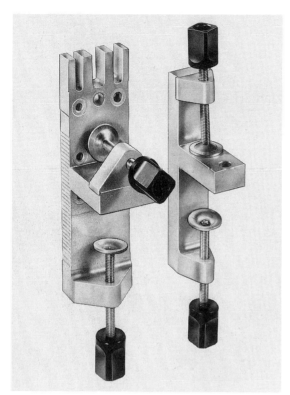

79 *The universal dowelling jig from Wolfcraft*

80 *The universal dowelling jig in use*

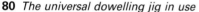

at one setting with this jig, to which the two components are secured with the built-in clamps – indeed the whole set-up can be fastened to the bench with the clamps as they are tightened on to the components. A spacer-clamp is also provided to ensure the components are properly aligned and held.

All the above jigs allow only for either a single hole, or a corresponding pair of holes, to be made at one setting, after which the jig has to be re-positioned on the material. Although with all these jigs there is a measure of automatic alignment, it is also necessary to carry out a fair amount of initial marking out and spacing for the centre of the holes.

The jigs described so far are designed for the hole that is in the edge or end of the material to be made in the centre of it, or very nearly so, and this position cannot be changed. The holes in the mating component are made to correspond, and thus for a corner joint the outer surfaces of the two parts finish flush, and this cannot be altered by any adjustment of the jig. Care is always needed to apply and use face marks, to use these as datum surfaces during the preparation of the joints, and to assemble the components with proper regard to them.

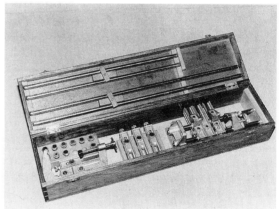

81 *The Record 148 professional dowelling jig*

The most adaptable and professional of all jigs for dowel jointing is the Record 148. The basic kit has guide rods which allow it to cope with wood up to 150mm (6in) wide, two bush carriers, and three pairs of bushes of $\frac{1}{4}$in, $\frac{5}{16}$in and $\frac{3}{8}$in diameter. Guide rods giving 305mm (12in) and 457mm (18in) capacity are also available, and additional bush carriers can be obtained. In addition, metric bushes of 6mm, 8mm and 10mm are produced.

This jig will cope with virtually every application where a dowel joint is suitable, and is readily adjustable in all directions. It works equally well on the face or end of the material, and can readily be used for

82 *Boring holes on face of component*

83 *The 148 in use on end of workpiece*

holes well away from the ends or edges of the wood. It adapts to wide boards or narrow framing material with equal ease. Once set and secured to the wood with its own clamp, all the holes needed in that component can be bored at one setting, assuming there are sufficient bush carriers on the rods.

The exception to the above is when the Record 148 is being used to prepare holes along the edges of two pieces of wood to be butt jointed and strengthened with dowels. For these joints, a little improvisation is required, with the jig being handheld while the holes are actually bored.

For the majority of applications, once the jig is set, only the minimum amount of marking out is needed for repeat joints to be made. For many uses, no initial marking out of any kind is required, apart from face marks. Again these must be used as datum surfaces. With some joints, all that is required is for a single line to be squared across the components.

Dovetailing jigs

Dovetail joints, as used for box-type constructions, provide a very strong method of assembly when properly prepared, and certain variations of these joints can be formed with the aid of a power drill and a jig. The jigs operate best on wood between 10mm ($\frac{3}{8}$in) and 25mm (1in), and normally cut a joint of the 'lapped' type. The jigs are generally designed for wood up to around 140mm ($5\frac{1}{2}$in) wide, but by careful repositioning of the jig, joints in wider material can be prepared. It is essential when using these jigs that the wood is first cut to exact length, and that the ends are dead square. The lap part of the joint must be allowed for when the length of the material is being established.

84 *Spiralux dovetailing jig*

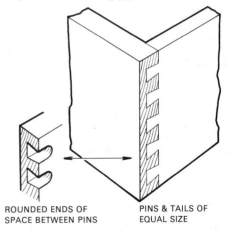

ROUNDED ENDS OF SPACE BETWEEN PINS

PINS & TAILS OF EQUAL SIZE

85 *Dovetails formed with jigs*

Although adjustments can be made to govern the tightness of the fit of the joint, the two parts that make up the dovetail, normally referred to as the 'pins' and the 'sockets', cannot be changed. Indeed, these parts are always of equal size, as the single cutter forms both pins and sockets. This means that some careful marking out is required initially to ensure that a part-pin does not occur at the edges, and because of this a

86 *Typical lap dovetail joint*

88 *Sockets formed with Wolfcraft jig*

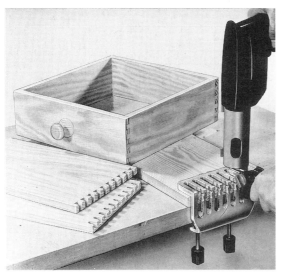

87 *The Wolfcraft dovetailing jig, preparing the pins*

final small amount of cutting might be needed with a dovetail saw.

The dovetailing jig made by Spiralux requires separate cramps to secure it to the wood. The one available from Wolfcraft has its own built-in cramps for holding it to the workpiece. The size and spacing are the same on both jigs, and also with both jigs a dovetailing cutter and a plain cutter are provided. The purpose of the plain cutter is to enable 'comb' joints to be made, the idea being that in chipboard there is less possibility of crumbling. In reality, neither the dovetail nor the comb joint is really suitable for chipboard. This material does not lend itself to cutting into small protruding parts. In addition, the cutters provided

with both jigs are HSS, and such cutters would have only a limited life when cutting in chipboard because of its abrasive nature. A speed of 3000 rpm is recommended for these cutters.

A final point must be mentioned regarding the joints produced by both these jigs. Because of the rotational cutting action of the cutters, the spaces between the pins have rounded bases, and in addition, this rounded end also incorporates the slope that is a continuation of the same slope to the sides of the pins. As the inner surface of the component carrying the sockets remains quite flat, this means that the fit of the joint on this inner surface is an imperfect one. For most applications of the joint this imperfection is of little consequence, but where maximum strength is required, it must be kept in mind that joints produced with these jigs have this minor weakness.

89 *Rounded waste to sockets*

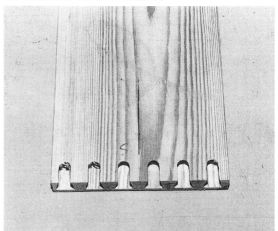

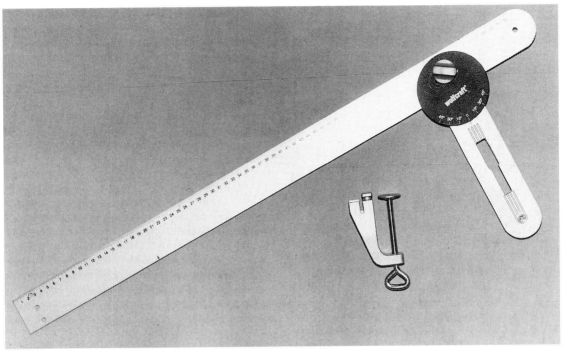

90 *The Wolfcraft Combinal*

Cutting aids

Wolfcraft produces a marking out aid which can also
be used as a guide for certain power tools. Known as a
Combinal, it is essentially a form of adjustable square
which can be secured to the wood by the clamp
provided. It is fully adjustable from 0–90 degrees, and
can be pre-set at a right angle, 15, 30 and 45 degrees.

6 || Circular Saws

The circular saw probably saves more physical effort than most other portable power tools, and in so doing is a great time-saver as well. A good saw properly used can cut timber so as to gain maximum economy, and is also sufficiently true and accurate as to require only the minimum of subsequent planing when a dressed surface is required. The circular saw can carry out far more than straightforward sawing, especially when used in the stationary mode set up as a saw bench.

91 *The Bosch PKS 54 circular saw*

The principal factor determining the size of a saw is the diameter of the blade. Allied very closely to this is the depth of cut. This is the amount of blade that projects from the sole plate of the tool and gives the maximum thickness of wood that can be cut. As with other power tools, the wattage of the motor is related to its power, and with a saw, it is particularly important for this to be adequate for the sawing to proceed at a reasonable speed when cutting at full depth in both hardwoods and softwoods. The rpm of the blade is a factor in both overall cutting efficiency and in speed of sawing, although the rpm is directly related to the diameter of the blade.

Speed

The critical speed of a circular saw is the speed at the periphery, that is the speed at which the teeth pass through the wood. The optimum speed for a circular saw is a peripheral rate of 9,000 feet per minute, or around 2,750 metres per minute. Thus for a blade with a diameter of 200mm (8in), which therefore has a circumference of around 630mm (25in), the speed should ideally be in the region of 4,300rpm. All these figures are approximate, and circular saws, small ones especially, will operate quite well at a speed much lower than the optimum. Little is achieved by having the blade rotating excessively fast, but if it is far too low then a slow rate of feed through the wood does not really compensate for this. Remember with portable power saws in particular there will be a drop in speed when the saw is operating under full load.

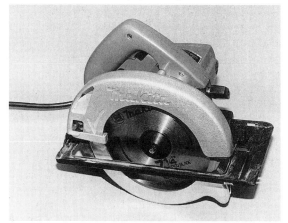

92 *Makita saw with 190mm (7¼in) blade*

Tooth forms

The saw blade, and especially the tooth size and form, vary considerably. The vast majority of saw blades in use at the present time have TCT teeth and this includes most of those fitted to portable saws. The alternative to this is the traditional one usually known as a plate steel blade. Tungsten carbide is exceptionally hard, and has a working life between sharpenings of between twenty and fifty times that of a plate steel blade. In addition, a TCT blade gives a very accurate and smooth cut, although this is dependent on tooth size and form, the diameter of the blade and other factors including, of course, the quality of the tool itself.

However, there is a price to pay for TCT blades. Their initial cost is quite high, and the tungsten is so hard that it requires special equipment to sharpen the blades, requiring them to be sent away for attention when blunt. In addition, like most materials that are particularly hard, tungsten is also brittle. This means that if a blade hits a foreign body in the wood, and particularly if the object itself is hard and is struck fairly suddenly, one or more of the tips can become chipped or even wrenched from the body of the tooth. Saw specialists can repair damaged blades by brazing new tips in place. While TCT blades are in no way intended for cutting mild steel, if a nail is accidently hit while cutting a piece of wood then the blunting effect is likely to be far less than it would be for a plate steel blade.

The blades fitted to most portable saws have the tooth pattern known as universal. This means that its shape and the way in which it is sharpened are designed for cutting both across and along the grain. In simple terms, a universal pattern tooth is of rip form in its outline with the leading edges given 'positive rake', but with the tops of the teeth sharpened to an alternative right and left bevel, which gives the teeth a slightly pointed tip either to the left or right. The rake of a tooth is the angle of the leading edge compared with a radial line drawn from the tip to the centre of the blade. A positive rake gives good ripping qualities, whereas the outer pointed sharpening of the teeth is an essential feature for smooth cross cutting. This is because the tips first sever the fibres of the grain before the body of the tooth forms the actual cut or kerf.

Blades intended solely for cutting with the grain have the tips of the teeth sharpened so that the top surface is square to the sides of the blade. This in turn produces a cutting edge which is square across its width without the characteristic point desirable for cross cutting. In addition, rip teeth have large gullets. The purpose of the gullet is to accommodate the sawdust while the tooth is in the kerf, and then expel this once clear of the wood. Large gullets mean that the teeth must be fairly well spaced, and blades intended for heavy ripping therefore have large teeth, not too close, and with deep gullets. While large rip

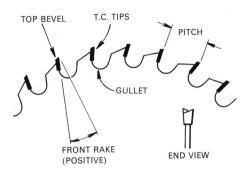

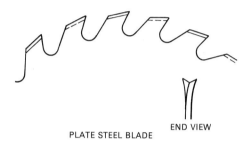

93 *Blades with combination teeth*

teeth are fast cutting and will produce an acceptably smooth surface, the quality of the cut is largely in proportion to the size of the teeth.

If a blade is intended to be used primarily for cross cutting, squaring, rebating, joints and similar cuts where the emphasis is on accuracy and smoothness, then smaller teeth are required. Smaller or 'finer' teeth mean more of them can be formed on a blade of given diameter, and the smaller gullets that inevitably result are still adequate as the rate of cutting, compared with a rip blade, is much slower. In addition, these blades usually have teeth with a negative rake, and are sharpened so that points are produced on the outer edges.

In broad terms, any pattern and size of tooth will cut any wood in any direction of the grain, but efficiency and quality are vastly improved when the teeth are matched to the nature of the cutting required. Rip teeth, and especially large ones, will produce a ragged cut and poor surface if used for cross cutting, and there is likely to be considerable splintering on the exit side of the wood. This is particularly true if cutting ply, or other man-made board with thin wood laminae on the faces. As well as the poor surfaces produced, large rip teeth, and especially blades of larger diameters, do not cut with quite the same accuracy as the alternatives.

In reality, saws are usually used to cut along the grain, and partly because of this the general practice of most manufacturers of portable saws is to fit blades with fairly large teeth as the standard blade supplied with the tool. The makers, though, are also under a certain amount of commercial pressure to offer a product that represents the best combination of price with quality. TCT blades are expensive, and therefore constitute a considerable proportion of the total cost of the tool. The cost of a blade is highly dependent on the number of teeth on it, and therefore rip blades are generally less expensive than blades with more but smaller teeth.

All manufacturers of portable saws offer a selection of blades as accessories to their products. For TCT blades, they will vary from around 12 teeth on a typical blade of 200mm (8in) diameter intended primarily for ripping, to possibly 40 teeth on blades produced essentially for cross cutting. Some economy blades are made with only around six teeth on them, making them more like 'cutters' than normal saw blades, but clearly such blades have limitations as to their use, and their life between sharpenings. Very few teeth on a blade make it unsuitable for smooth cutting of thin wood, and also for ply and similar boards. The fewer the teeth, the more actual work each tooth is carrying out as the sawing proceeds, and therefore such a blade will require sharpening more frequently than a blade with more teeth on its periphery.

94 *Typical saw blades*

It is good practice to use a blade with sufficient teeth so that as the wood is being sawn there are always at least two teeth in contact with the wood. When this is not the case, the cutting action, as well as the surface produced, becomes ragged, especially if

the tool has a speed a little lower than the ideal. Because of the circular shape of the blade, the teeth as they pass through the wood form a kerf, the leading edge of which is an arc. The distance around the arc is greater than the thickness of the wood, and the less the projection of the blade through the sole of the tool, the greater the extent of the arc. What this means is that by adjusting the tool so that the blade will only just penetrate the thickness of the wood being sawn, then the larger the extent of the arc of the kerf and therefore the larger the number of teeth in contact with the wood. Thus limiting the projection of the blade can compensate in part for the size of the teeth, and lead to smoother cutting.

Blade types

Most blades intended for use in a portable saw tend to be a little thinner than a corresponding blade for mounting in a saw bench. As well as the blades offered by the makers of the tools, most of the saw blade specialists produce blades designed for use in a portable saw. Indeed most of the tool manufacturers will have the blades made for them by firms whose primary business is to produce blades and cutters. Most blades are offered in different qualities such as standard and premium, the differences including the number of teeth, the actual size of the TCT, and the thickness of the blade measured by the width of kerf produced. For small blades of the diameters used in portable saws, the thickness can vary from around 2.6mm to 3.5mm ($\frac{3}{32}$in to $\frac{1}{8}$in). The advantages of a thinner blade are that the cutting absorbs less energy, which means less strain and slowing of the motor, and less waste of wood from a thinner kerf. The disadvantage is that a thinner blade is also less stiff, can

95 *Black and Decker 'Plus' showing rip fence*

therefore deflect slightly, and is likely to be a little less accurate because of this. Some blades fitted to portable saws are coated in a non-stick material, the idea being that this reduces friction and therefore the loss of energy. Where a saw is to be used extensively for cutting ply and similar products a blade specially intended for such materials should be fitted. Opinions vary as to the best shape for the tips of the teeth on blades for man-made boards, and there is ongoing development. Remember, though, that a top quality blade of any pattern must be matched to a tool of equal quality if maximum performance is to be achieved.

Features of power saws

Because of the widespread use of saws throughout all branches of woodworking, the power tool manufacturers each produce a wide range of these products. In general terms, they all have similar features, the

96 *Lightweight model from the Black and Decker range*

97 *This Skil saw has a depth of cut of 65mm (2½in)*

essential difference being size combined with robustness. The smallest saw made has a blade diameter of 125mm (5in), a maximum depth of cut of 40mm (1 9/16 in), a motor rated with an input of 550 watts, and a no-load speed of 3,500 rpm. At the other end of the scale is a saw with a blade diameter of 335mm (13¼in) giving a depth of cut of 128mm (5in). This is powered by a motor rated at 2,000 watts input, and has a no-load speed of 3,200 rpm. The most popular sizes are those with a blade diameter of around 200mm (8in), which gives a depth of cut of approximately 65mm (2½in).

Sole plates

Saws have the facility for the sole plate to be pivoted, which results in an angle other than 90 degrees being created between the sole and the blade. This is to enable bevel cuts to be made at any angle between a normal square cut and 45 degrees, which is the limit of tilt. The sole plate is locked at the required angle either by a thumbscrew, a quick acting lever, or a combination of both, and a scale is provided as a guide to the setting. This scale, and others usually marked for the depth of cut, and on the fence for ripping, should only be taken as a guide, and in all cases where close accuracy is wanted, a trial cut should be taken and checks made that the cut is exactly as wanted. Some, but not all, saws have an adjustable zero screw so that when the sole plate is returned to its normal setting, a 90 degree angle is automatically re-established.

98 *Adjustable sole plate of the Elu 182*

When angle cuts are made, the effective depth of cut is reduced, the loss of cutting depth depending on the angle. When the sole plate is set for the maximum

99 *Depth of cut control on the Bosch*

100 *Guard retracted showing small depth of cut*

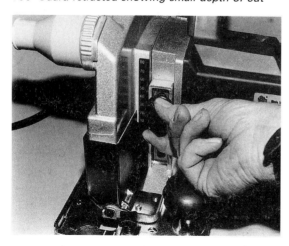

101 *Depth of cut adjustment on the Elu*

bevel angle of 45 degrees, the resulting depth of cut is approximately 70 per cent of what it is for square cutting.

The depth of cut, whether for square or bevel sawing, is governed by the amount of blade protruding through the sole. This can be adjusted, and two alternative systems are in use. The slightly simpler arrangements allows for the sole plate to be pivoted, the pivot being at the front and the adjustment made at the rear. With this adjustment at its greatest, which gives the minimum depth of cut, the relative positions of the handles are changed but cutting can still proceed with the handling of the tool remaining comfortable. On the second system of adjusting the depth of cut, the sole plate remains parallel to its original position by lugs sliding along slots positioned at the front and rear. This is a very solid arrangement, and has the advantage that the positions of the handles relative to the surface of the wood remain unchanged.

Riving knifes

A key feature of any circular saw is the riving knife. Its purpose is to ensure that the kerf does not close in as the wood is being sawn. There is often a natural tendency of the wood to 'bow' slightly on the newly sawn surfaces, resulting in the wood on both sides of the kerf moving together. This is because the equilibrium of the wood is thrown out of balance by the sawing, the extent of this depending mainly on the dryness of the wood and the uniformity of this, and its cross-sectional size. If the kerf should close in, the wood can grip the blade at the rear part of the saw, that is as the blade emerges from the sole plate. This would result in the wood being forced off the tool, and can be highly dangerous. There is a second danger from the wood closing in on the blade in the absence of a riving

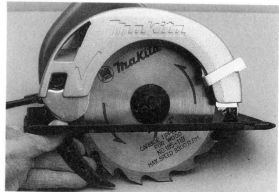

102 *Guard and riving knife on the Makita*

knife, or one which is too thin. Excessive friction on the blade because of lack of proper clearance can cause overheating, and this in turn can lead to buckling of the blade. A distorted blade will not run true, and therefore cannot cut with any degree of accuracy.

The thickness of the riving knife should be as close as possible to the gauge of the blade, or very slightly thicker if anything. Thus if the kerf does close it will first touch the teeth, which will simply cut the wood, then come into contact with the riving knife, which will keep it from gripping the body of the blade. For the riving knife to be fully effective, the inner edge should follow the curvature of the blade, and be positioned between 3mm and 10mm ($\frac{1}{8}$in and $\frac{3}{8}$in) from it. It must be positively anchored in position so that it is exactly central with the line of the blade. The leading edge should be thinned slightly to allow for easy entry into the kerf, and the extent of the knife should be such as to be slightly within the limits of the teeth when viewed from the side. This is to allow rebating, grooving and similar cuts made less than the full thickness of the wood to take place with the riving knife still in place.

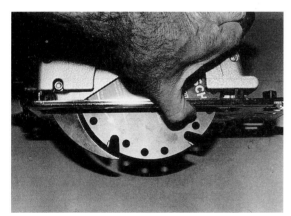

103 *Riving knife must be close to and follow curvature of blade*

The riving knives on small saws are themselves fairly small and lightweight, even though quite adequate. They can, though, become bent a little out of place, and care should be taken to ensure that everything about the knife is correct for it to be fully effective. For this reason, when a replacement blade is fitted its thickness or gauge should be checked against that of the knife. In addition, a blade should not be fitted which is much smaller in diameter than the original, because of the necesity of keeping the gap between the knife and the blade within the limits that are stated.

Guards

Just as important as the riving knife from a safety point of view is the guard. Those fitted to portable saws are spring-loaded, are pushed out of the way by the wood as the sawing takes place, and automatically return to enclose the blade at the end of the cut. Some guards have small wheels fitted at the forward end to enable them to ride up more easily on the corner of the wood as cutting commences. Most guards on portable saws are of the split type, with a slot down the centre to allow each half to straddle the riving knife. Not only does the guard give protection directly to the user, it also prevents the accidental cutting of the lead, particularly at the end of operations as the blade tends to continue running for several seconds once removed at the end of a cut. Additionally, the guard protects the blade from damage, especially if the tool is put down while still running.

Arbors

Arbor sizes vary from saw to saw. Most are metric and range from 16mm to 30mm. It is essential that the hole diameter of the blade corresponds exactly with the arbor, and fitting close approximations of imperial bores to metric arbors, and vice versa, would lead to severe problems as the blade would be running eccentrically. This would create a state of imbalance that would impose a strain on the bearings. In addition, those teeth at the greatest distance from the axis of rotation would be doing most work, and the cutting would therefore be erratic. It is essential for the proper functioning of the saw that the blade is fully in balance, and that all the teeth project by a uniform amount and thus perform an equal share of the cutting.

However, in certain cases it is possible to fit a blade to an arbor where the diameters of the hole and the arbor are not the same. A reduction ring is used to make up the difference between the two sizes, and these can be supplied in varying combinations of inner and outer diameters by the saw specialists. Only perfect fits should be regarded as being acceptable. It should be noted, however, that the blade does not always fit directly onto the arbor but sits on a step on the inner flange, and therefore the diameter of this step becomes the critical dimension.

Cleanliness around the arbor, the inner and outer flanges, is essential, and these parts should be wiped clean whenever the blade is removed. Dirt, especially on the inner flange, can cause the blade to run out of true. The outer flange is usually held onto the arbor by either a machine screw, which requires a spanner, or a socket screw, which requires an Allen key. The threads are of the normal right-hand type. To slacken the screw, the blade has to be locked to prevent rotation, and built-in blade locks are not normally provided for reasons of safety. Some blades have holes in them and a hole in the body of the saw, and inserting a steel pin or the shank of a small Allen key into the corresponding holes prevents the blade from rotating. In the absence of holes, a piece of wood should be held against the teeth to prevent rotation. When a blade is refitted, the screw should be tightened simply by holding the blade in the hand. Only moderate tightening is required, as the rotation of the blade has a self-tightening effect.

Wobble sawing

A small number of manufacturers supply special sets of washers to allow 'wobble sawing' to take place. Wobble washers vary slightly in the way they work. Most incorporate two of these washers with matching sloping faces that grip the blade in the out-of-true position. By rotating the washers on the arbor, settings between zero and the maximum can be made, the greatest setting for this type of saw usually being around 10mm ($\frac{3}{8}$in).

The main purpose of wobble sawing is to form a cut that is much wider than the thickness of the blade, and with a width which can be controlled. It is particularly useful for forming grooves and trenches, and especially where these have to be formed on a repetitive basis. Although wobble washers are at their best when the width of the cut required is within their capacity, cuts greater than the maximum setting can be made by taking multiple passes. This saves time in comparison with multiple pass grooving without the use of washers. Wobble washers can also be employed for rebating and similar cuts where the amount of waste to be removed is greater than the thickness of the blade.

Wobble sawing, even at its best, suffers from an imperfection. Because of the way in which the blade oscillates as it rotates, the outline of the tips of the teeth when in line with the blade is of an arc, the radius of which equals the radius of the blade. The result of this is that the bottom surface of the kerf, as formed when grooving, for instance, is not quite flat, but slightly concave. In addition, a fairly thin blade can behave in a slightly curious way when mounted between wobble washers. During rotation, the blade is subjected to a centrifugal force. The effect of this is to influence the blade to follow a flat path. The path of

the blade can only flatten if the blade itself flexes a little, and for this reason a blade intended for use with wobble washers should be of a heavy gauge and therefore quite rigid. In addition, note that a smaller blade is less likely to flex than a larger one.

When cuts are being made with wobble washers in use, the feed rate of the wood – or the saw – should be kept fairly slow. A high feed rate, or an erratic feed rate, can result in the width of cut being less than it should be, or varying a little. This is because the blade wants to take the line of least resistance, especially when it is being made to function at a high work rate. The line of least resistance is the line that removes least wood – the resistance the wood offers to being cut can cause the blade to flex slightly inwards, hence the narrowing of the cut. The riving knife is ineffective when wobble washers are fitted, but the need for it is removed by the nature of wobble sawing – the kerf so made is wider than the thickness of the blade.

Although wobble sawing can be carried out with the saw used in the hand-held mode, the technique lends itself more readily to use in the fixed, inverted position in a table designed for converting a portable saw into what is effectively a saw bench, or saw table.

Switches

The switch on saws is of the trigger type, but to prevent unintentional running of the tool when not required, lock-on switches are not fitted. Indeed, as an additional safety precaution, at least one manufacturer fits a catch to the switch, and this has first to be depressed before the trigger can be operated. All switches are the positive on-off type. There is no need for a saw to be fitted with variable speed, and the blades are not equipped with any breaking system. Most saw bodies are designed so that the dust created is deflected through an exhaust at the side and clear of the operator, with the option of linking up to a vacuum extractor. A special adaptor piece is usually required to make the connection to such an extractor.

Rip sawing

The operation most performed by circular saws is that of ripping, with the cut being made parallel to the edges. Because of this, a rip fence is normally provided as standard equipment. This fits into the front end of the sole plate and its adjustment is controlled by thumbscrew or knob. The fence only normally operates on the right-hand side of the blade. This ensures maximum contact and therefore support

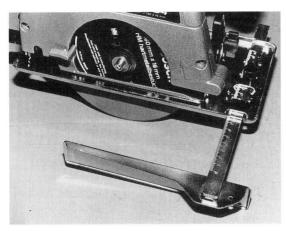

104 *Graduated rip fence on the Bosch*

of the sole plate on the wood. Graduations on the arm of the fence should only be taken as approximate. It is far better to make the setting with the use of a rule, and better still for critical work to check the width being sawn after an initial start to the cut just sufficiently long to allow a measurement to be made. Straight sawing is not automatically made, although it should not be possible to cut the material wider than intended. However, the saw can wander to the right resulting in a cut narrower than the setting made, and the operator must consciously control the saw so that the fence is tight, and in permanent contact with, the edge of the wood.

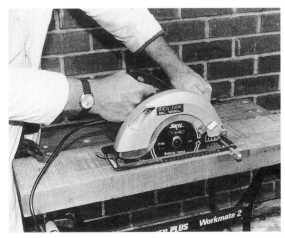

105 *Ripsawing using the fence*

The above assumes that the original edge of the wood is straight or reasonably so. If not, then any curvature will be repeated in the cut as the fence provides a parallel cut. Curved edges will not be critical when the pieces being ripped are of small section, as such pieces are likely to be flexible anyway.

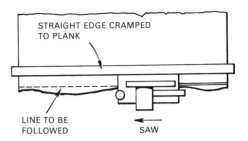

106 *Trueing one edge of waney plank*

Not all sawing is required to have the same degree of accuracy but, where this is important, one edge of a distorted board or plank must be trued up at the outset. This can be carried out by plane, or by using the saw against a guide batten known to have straight edges. The batten must be cramped to the material, and, of course, only the minimum required to produce a true edge need be removed.

Because the fence of the saw will only operate up to a limited distance from the blade, it cannot be used for all ripping operations. For cutting wide boards, plywood and similar material, the use of a straight edge as described above can be adopted. Where a taper cut is required then, whether large or small, the fence cannot be used. Again, such cuts can be readily carried out by using a straight edge, and a commercially produced one is available for use with portable power tools. This straight edge is an aluminium extrusion, will cope with material up to 2,440mm (96in) in length, and incorporates two cramps that allow it to be used over any length up to its maximum.

107 *Ripsawing using a straight edge*

The saw can be used in an entirely freehand way, and for many operations this provides the best option – initial cutting of rough planks, for instance. As even 'rough' sawing usually needs to be straight, a line must first be made on the wood showing where the

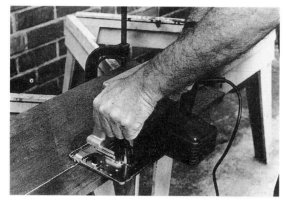

108 *Freehand sawing following a line*

109 *Forming groove on edge of workpiece*

cut is to be made. On the front of the sole plate of the saw is either a notch or a mark, made in alignment with the blade. Thus by keeping this mark directly over the line on the wood, the blade will follow the path required. With a little practice, this method of working is quite accurate, even though the blade itself cannot be seen during the cutting.

Cross cutting

Cross cutting techniques vary a little from those of ripping. Despite the fact that many of the manufacturers depict in their promotional and packing material illustrations showing the cross cutting of boards and planks taking place with the fence guiding the saw, this is not a method of working to be recommended. First, it assumes that the end of the plank is square to the edges, and there is far more possibility of this not being the case than there is that the ends are always at 90 degrees. Any inaccuracy of the original end would be largely repeated in cuts made using this end as a guide. Secondly, the extent of the end is usually very

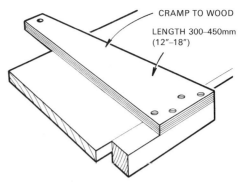

CRAMP TO WOOD

LENGTH 300–450mm (12"–18")

110 *Simple cross cut guide*

limited, and therefore offers very limited guidance and control to the fence and thus the saw. Once a board or plank has been ripped into narrower pieces, then cross cutting using the fence becomes almost impossible. In any case, the need for short lengths of material, of the size that could be cut using the fence, are not likely to be great.

111 *Cross cutting using home-made guide*

112 *The Combinal makes an ideal guide*

Cross cutting of big material is best done with a guide secured to the wood. This can be in the form of a simple home-made jig set accurately at 90 degrees and cramped to the wood as required. An alternative to this is the Wolfcraft Combinal. This can be set to a right angle, or any angle between this and 0 degrees. Although one end of this device incorporates a cramp for securing to the wood, it is advisable when using on wide wood, to add a second cramp on the free end of the arm.

Supporting work

It is essential when using a circular saw in the handheld mode that the material being cut is properly supported. The height should be such as to allow comfortable working, and over-reaching must be avoided. Trestles or similar devices with a height of around 600mm (24in) provide good support, but positioning is important. They must be clear of the line of sawing to prevent damage to them, and in addition must offer proper support after the completion of the cut, as well as before.

Stationary use

Several of the power tool manufacturers produce their own tables which a portable saw can be used in conjunction with, thus converting the tool into a stationary saw bench. Such a set-up adds considerably to the versatility of a circular saw and is especially useful for cutting small sizes of material, and for forming certain joints. The AEG table has provision, with appropriate extra equipment, for box combing to be carried out, while the Elu table will operate with the

114 *The Elu combination table*

115 *The Elu table set up as a snip saw*

saw mounted below the table, or on an arm above it when it can be used as a snip-off saw. In addition, the Elu table can be used for mounting a router in the inverted mode.

What must be realized is that these saw tables will accept only a limited number of the saws from the same manufacturer, and that there is not complete compatability across all the range. They are not designed to accept saws from other manufacturers, although it is possible that some may be safely mounted. Partly because of the inaccessibility of the switch on the saw – or router – when mounted for static use, these tables provide a separate switch mounted on one of the legs, this control also incorporating a socket into which the plug of the tool is

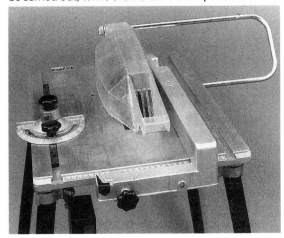

113 *The AEG saw table*

116 *Underside of AEG table showing mounting of saw*

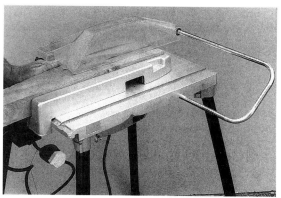

119 *Ripping on the AEG table*

The guard attached to the saw cannot be used when it is mounted in a table, and therefore a separate guard secured to the table is needed. Arrangements for the riving knife vary. On some tables the knife of the saw is used; on others a separate one is required. Note that there is a slight loss of cutting depth when the saw is used secured to the table because of the thickness of the table itself combined with the method of securing the tool.

117 *The independent switch on the Elu*

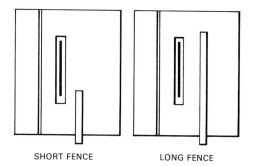

SHORT FENCE   LONG FENCE

120 *Plans of tables with guards removed*

Because ripping is the commonest operation carried out on a bench saw, a principal part of the table is the rip fence, normally positioned to the right of the blade. It is essential for the fence to be parallel with the blade, and this must be checked when the saw is secured to the table. Opinions vary as to the exact length of the fence in relation to the blade. Some users prefer what is referred to as a short fence, that is a fence that extends just beyond the teeth on the near side of the blade; others are happier with a long fence extending a little past the rear of the blade as it emerges from the table. Generally, for sawing fairly thick wood, a shorter fence has advantages as it removes the restriction alongside the blade during actual sawing and allows the sawn piece to flex away

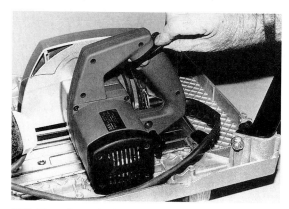

118 *The trigger lock is essential for inverted use*

connected. The switches are usually of the no-volt release type to prevent unintentional starting. Because of the switches on saws not being of the lock-on type, either a plastic clip or strap must be used in order to keep the trigger in the depressed position.

from the blade if it has a tendency to do so. For sawing thin material and man-made boards a long fence provides maximum support. A short fence can also be used as a 'stop' when cross cutting small pieces, but a long fence is essential when cuts are being made only part way through the wood, such as when grooving or rebating.

The fences supplied with the tables are of the long type, and therefore if a shorter fence is preferred then this has to be achieved by adding a piece of wood extending only part way along the existing fence. Holes may have to be drilled, or drilled and tapped, for this, but adapting a fence to suit the way in which the saw is used is not uncommon. Most fences are usually fairly shallow, with the height less than the amount by which the blade penetrates the table. On the odd occasions where the thickness of wood to be cut is greater than the capacity of the saw, this can be carried out by cutting part way from one side of the wood, then reversing the material for a second cut from the opposite side. This is sometimes known as deep sawing, and it is essential to have the wood properly supported throughout the sawing. This can be achieved by adding a false fence of suitable height, but it is essential for this to be of the long type extending beyond the rear of the saw.

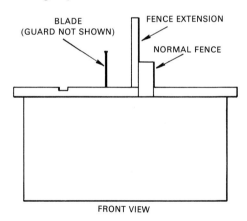

121 *Heightening normal fence*

Safety

Whenever making any cut along the grain, whether it be normal ripping or a cut made part way through the wood, it is essential that a properly prepared 'push stick' be used as the end of the wood approaches and passes the blade. Never attempt to remove any piece of wood or offcut from the vicinity of the blade while it is still running, and should a piece of waste disappear part way down the gap alongside the blade and the

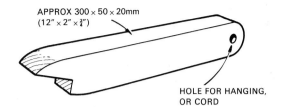

APPROX 300 × 50 × 20mm (12″ × 2″ × ¾″)

HOLE FOR HANGING, OR CORD

122 *Push stick*

table insert, the saw must be stopped before the offending piece is withdrawn.

It is always easier to hold and control a fairly long or heavy piece of wood at the start of the cut than at the completion. In addition, for most cuts, only one piece of wood has to be controlled at the commencement, whereas this becomes two as the sawing is completed. Because of this, assistance should always be sought when cutting material of a size or weight that might cause the operator to over-reach, lose control or balance, or in any way create a risk while sawing.

For cuts that are made only part way through the wood the guard will have to be removed in most cases. However, even if the guard has to be positioned high over the blade during such cuts then it should be fitted. Its use draws the attention of the operator to the danger area below. Normally, a guard should be fixed so that its lower edge overlaps the highest teeth. When cuts are made at less than the full thickness of the wood, the wood itself completely covers the blade and therefore guards it. Special care, and, of course, the use of the push stick, must be used at the end of such cuts, as the blade reappears at the end of the wood without warning.

Rebates

Rebates are normally formed by taking cuts from two adjacent surfaces until they meet. When rebating, the usual practice is to have the edge to be cut against the fence, as this give the maximum accuracy to the rebate. Where the rebate is particularly large relative to the sectional size of the wood, the opposite method of working should be followed so that the fullest support from the fence is gained.

Grooves

Grooves have to be made by taking a series of cuts alongside the first, so as to widen the initial kerf to the width of groove required. This is referred to as taking a series of 'passes', and with any operation where

123 *Forming a rebate on the saw table*

adjustments to the set up are required to complete the cutting, it is highly desirable that all pieces of wood involved should be similarly treated as each adjustment is made. This ensures uniformity, as well as saving time. Where the edge of a narrow piece is being grooved, a high, long fence is required.

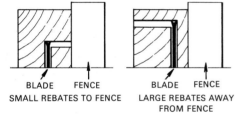

BLADE   FENCE   BLADE   FENCE
SMALL REBATES TO FENCE   LARGE REBATES AWAY
                              FROM FENCE

124 *Sawing large and small rebates*

Bevel ripping

Bevel ripping can be carried out on a saw table by using the tilting facilities of the tool. For this, though, once the blade has moved a few degrees from the vertical, the 'bevel ripping' table insert has to be

125 *Special table insert required for bevelled sawing*

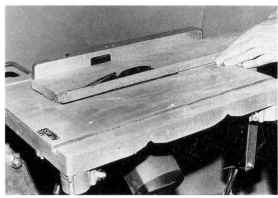

126 *Bevel ripping, guard removed*

brought into use. This replaces the usual table insert through which the blade protrudes, its wider slot allowing for the blade to tilt.

Cross cutting

For cross cutting, a cross-cut fence, also known as a mitre fence or slide, must be used. This is designed to slide in a groove, usually on the left-hand side of the

127 *Cross cutting using mitre fence*

blade, although some tables including both the AEG and the Elu have grooves machined on both sides of the blade. Cross-cut fences can be adjusted so that angles between 90 degrees and 45 degrees can be formed, but once again, the protractor scale must be taken as an approximate guide only. Note that a mitre cut is a cut made at an angle other than 90 degrees (but not necessarily 45 degrees) to the *edge* of the wood, whereas a bevel cut is made at an angle other than 90 degrees to the *face* of the wood.

On both the AEG and Elu saw tables, the mitre fences allow for the front face to be adjusted

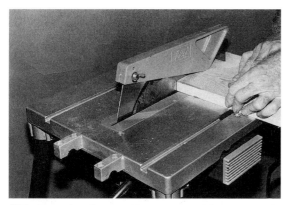

128 *The mitre fence will operate to left or right of blade*

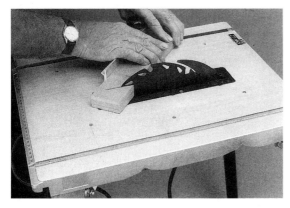

129 *Any angle from 45 degrees to 90 degrees can be sawn*

sideways. This is a refinement not always seen on this type of equipment. The main purpose of this is to ensure that the fence can be adjusted so that, whatever the angle being sawn, support can be given to the wood close up to the point of cutting.

When cross cutting, the wood should be held tight against the mitre fence with both hands and moved steadily into the revolving blade. The piece being sawn off must not be moved forward with the hand, as this could result in the kerf closing in on the blade with a risk of binding. The long fence must not be used as a length stop when cross cutting, or the cut piece can become wedged between the blade and the fence. If the sawn piece should twist slightly, it can exert such pressure on the blade as to buckle it, with serious consequences. For the same reasons, cross cutting without the mitre fence, but simply holding the end of the wood against the rip fence for guidance, should never be attempted.

When a lot of cross cutting is carried out it is worth adding a false face of wood to the mitre fence. This not only provides additional support to the material being sawn, it enables stops to be easily fixed to this face. By keeping the projection of the blade to the minimum required for the work being tackled, the fence can straddle the blade and thus offer support throughout the cut. This technique also offers the opportunity of adding stops either to the left- or right-hand side, or, of course, pencil marks can be made to aid the cross cutting as alternatives to stops.

Cutting joints

One of the commonest joints used in many branches of woodworking is the mortise and tenon, and the circular saw provides one way of forming the tenons. Assuming the tenon is positioned in the centre of the thickness of the wood, the height of the blade above the table must be adjusted so that it will cut into the cheek of the tenon just up to the gauge line. Either by using a short rip fence, or a stop on the cross-cut fence, the set up is adjusted so that the blade will initially cut the shoulder of the tenon. By moving the wood fractionally to the left as each cut or pass is made, the whole of the waste from the tenon is sawn away, and the operation repeated on the opposite side of the wood. When the tenon is off-centre, the height of the blade will need adjusting when the wood is

130 *Forming tenon on Elu table*

reversed, and for tenons with 'long-and-short' shoulders, the stop will require resetting. When working from the end, as is the case when stops are used, all similar components must be sawn to identical length, and with the ends dead square. Alternatively, neither stops nor the fence need be used; simply judge by eye whether the cutting is exactly in line with the shoulder of the tenon. It is always advisable to make the first kerf adjacent to the shoulder. Working from the end inwards can result in a 'part cut' being required to complete the cutting. This often leads to inaccuracy as the blade may drift slightly.

Using basically the same techniques as outlined above, simple half-lap joints can also be cut, as well as rebates across the ends of the wood as would be used in a rebated box corner joint. Likewise, cuts can also be made away from the ends of the wood to form trenches as might be required as a part of a half-lap joint, or as one half of a housing joint. By combining these methods of working, other joints such as the corner tongue and groove can be made entirely on the saw bench. An appropriate blade, as described earlier in this chapter, must be fitted to the saw when joints, mitres, and similar cuts demanding accuracy and good finish are being made. In addition, when forming joints made by cutting from both or all four surfaces of the wood, it is essential that all the material is accurately prepared to exact width and thickness. Any variations in the sectional dimensions of the material will be reflected in the accuracy of the joints cut, with a consequent inconsistency in the quality of the fit.

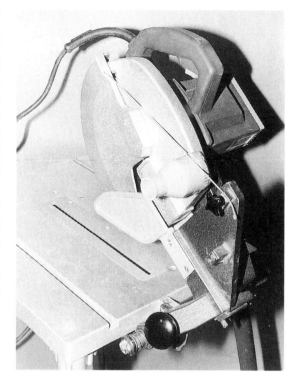

131 *Special bracket supports saw above table*

Compound cuts

By having the blade tilted at an angle to the table, and the mitre fence set to other than 90 degrees a cut that is both bevelled and mitred can be made; this is called a compound cut. A full range of compound cuts cannot be made, though. While the mitre fence can be set to the left or right to give the mitre cut, the blade can only be made to tilt to the left. Where compound cuts have to be made so as to fit together and 'match' on wood of plain section one piece must be reversed on the table for this to be achieved. Matching compound cuts cannot be made on wood that is moulded, rebated, grooved, or otherwise shaped so that it cannot simply be reversed. This is not just a limitation of the equipment being discussed; it applies to most static saw benches even when quite sophisticated. Compound cuts, though, are rarely required.

Snip-off adaptation

The Elu saw table can be adapted by the addition of extra components into a snip-off saw, the tool being mounted in a pivoting arm above the rear of the table. Note that the tool and the arm must be properly matched for this function. There is an adjustable spring in the arm to control the tension so that the arm and saw are automatically raised after the cut, and remain in the higher position when sawing is not

taking place. A small cable connects the guard of the saw to the arm, so that as the saw is lowered to make the cut, the guard swings out of the way and yet maximum protection of the blade is retained. The mitre fence must be used in conjunction with the snip saw, but it must be fixed to the rear of the table. It can be readily locked into either of the two grooves on the table, and the face of the fence can be adjusted as it is in normal use to give mitre cuts to the left or right. The use of the handle and switch is retained in this mode of operation.

The principal advantage of the snip-off saw, compared with using the normal cross cutting facilities of the saw table, is that far longer and heavier pieces of timber can be tackled. When the saw is mounted beneath the table, it is the wood that has to be moved and therefore there are limitations as to the sizes that can be physically controlled and moved safely forward into the blade. In the snip-off mode, the saw is lowered on to the stationary wood, and thus the restrictions of length are largely eliminated. However, for regular cutting of long pieces, supplementary tables need to be built and positioned on either side of the saw table, their heights being equal. Thus adequate support is given to the timber being sawn, not only just before the cutting, but more importantly after

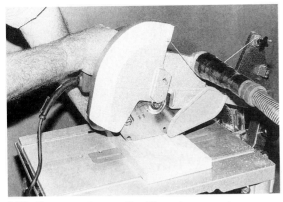

132 *Cross cutting on the Elu snip saw set up*

the wood has been cross cut and the balance of the material has been changed.

Because the saw pivots downwards but in an otherwise fixed way, the cross-sectional size of timber that can be cut by the snip-off saw set up is limited to around 100mm by 50mm (4in × 2in). In addition, the width of the wood that can be tackled is reduced when mitre cuts are made. The snip-off saw cannot be used for forming tenons and similar joints.

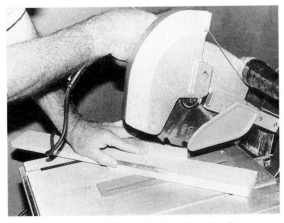

133 *Mitre cuts readily made by adjusting fence*

Box combing

The box comb joint has come back into favour in recent years. One reason for this is the development of combing jigs for use with circular saws. The jig is used in conjunction with wobble washers, set so that a cut from the blade produces a kerf usually between 6mm and 10mm ($\frac{1}{4}$in and $\frac{3}{8}$in). Careful setting out of the joint is essential to ensure that the width of the wood is equally divided into slots and projections, or a partial cut will be formed at the final edge of the component.

The AEG combing jig is designed to be secured to the mitre fence, and has a single adjustable finger that engages in one slot as the next is cut. Working in careful sequence is essential when box combing, and instructions relating to the jig must be followed exactly. Any width of wood can be cut on the jig, but the maximum thickness that can be tackled is 20mm ($\frac{3}{4}$in), adequate for most applications. While theoretically any length of wood can be box combed on the jig, in practice this is limited by the support offered by the jig itself. The wood is handheld as it is fed past the blade, and therefore a safe balance has to be maintained as the cut is formed. Wood up to about 450mm (18in) can be jointed on this jig, providing it is not also particularly wide and thick, and therefore heavy and difficult to hold upright against the limited height of the face of the jig.

7 || Jig Saws

Although a jig saw will make both straight and curved cuts, it is for the latter that it is especially designed and intended. Because of its reciprocating, narrow blade, it negotiates curves with ease. The static machine used for similar curved sawing is the bandsaw, but the portable jig saw has at least a couple of positive advantages over even the best of bandsaws. A bandsaw cannot make a cut that is completely internal because of the continuous nature of the blade, and it can operate at only a certain distance from the edge of the wood because of the restrictions of the frame of the machine. The jig saw is particularly useful for internal cut-outs, for example as required for a recessed sink in a kitchen worktop, and its use is just as advantageous in these cases whether the cutting is straight or curved. The jig saw can be used at any distance from the edge of the workpiece, and on very large material following the line is much easier than when bandsawing as it is the tool that is moved and not the wood.

The jig saw cannot be regarded as a serious alternative to the circular saw for straight sawing, whether this be ripping or cross cutting. Accuracy and rate of cutting are lower than for a circular saw. The cutting speed of a particular jig saw is very much related to the thickness of the wood and therefore when it is cutting at or near its maximum capacity the speed of progress is reduced. The jig saw does lend itself more readily than the circular saw to cutting materials other than wood. A wide range of blades is available for jig saws, relatively low in price and specially prepared in size, outline, hardness, temper and teeth profile for cutting most common materials including ferrous and non-ferrous metals, plastics, glass fibre, and most building boards.

Blades

Blades intended primarily for wood have two main patterns of teeth. For quick cutting, and especially for sawing up to the maximum capacity of heavy-duty

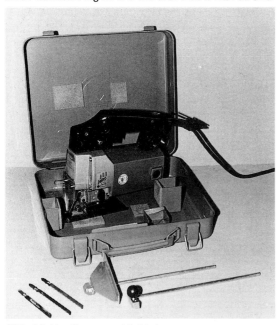

134 *Makita jig saw with kit box*

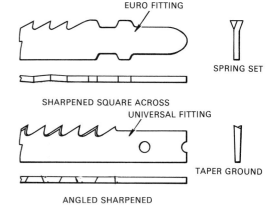

135 *Jig saw blades, different patterns, different fittings*

tools, which is usually around 60mm (2¼in), the teeth are sharpened square across in a similar way to a traditional hand rip saw. The quality of finish from such a blade will be fairly coarse, and there is high probability of some splintering of the surface when cutting across the grain.

For producing a cleaner cut a blade with the teeth sharpened at an angle so as to create an outer tip higher than the rest of the tooth is better. As well as the cut from this pattern of teeth being smoother, there is far less splintering on the top surface when cross cutting; this is because of the way the pointed tips sever the fibres of the grain. Blades with angled teeth should be used on ply and other man-made boards, as most grades of these boards are prone to surface splintering.

Regardless of their pattern, the size of the teeth affects the cutting action. The larger the teeth, the quicker the sawing, but the coarser the surface produced. Thus a blade with large teeth is more suitable for sawing thick material providing the tool can cope. The governing factor for maximum capacity is the jig saw itself, and not the blade. For the smoothest cutting a blade with small teeth will give the best results, but with such a blade the rate of sawing will be proportionally slower, and possibly tediously slow in thicker materials. Largely because of this, the blade makers usually produce their blades so that, broadly speaking, the length of the blade is related to the size of the teeth – typical working lengths across a range being from around 50mm (2in) to 85mm (3⅜in). When a blade is selected, the length must be such that it penetrates the wood at the highest point on the upwards stroke.

The size of teeth on these blades is measured by the pitch. This is most easily interpreted as being the distance from the tip of one tooth to the tip of another. This can range from 1.2mm to 4mm (1/16in to 3/16in), with teeth as fine as 0.7mm (1/32in) for metals. What has to be remembered regarding size of teeth is that the space between the tips – the gullets – is used to carry away the sawdust. Thus a balance always has to be kept between size of teeth and rate of cutting on the one hand, and clearance of sawdust via the gullet on the other.

Set on the teeth can be of different types, including the traditional bending outwards of alternate teeth which is known both as side set and spring set. This method of providing set is particularly suitable for the larger teeth that are sharpened square across, that is, for fairly heavy duty sawing. For finer work in thinner materials, where teeth sharpened at an angle are preferred, the set is often produced by taper grinding the thickness of the blade so as to make the rear edge

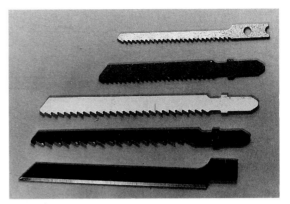

136 *Selection of jig saw blades*

thinner than the teeth, the whole purpose of set being to create clearance in the kerf and thus prevent building. Some blades adopt a combination of side set and taper grinding. A third method of creating set is to make the whole of the leading edge of the blade slightly wavy along its length. This is almost always adopted for metal cutting blades, but it also used on blades with relatively small teeth and intended particularly for ply and other board material.

As well as the blades described above, there are a small number of more specialized types, some for wood and some for other materials. There is a TCT blade, able to tackle wood, hard plastics, and sheet materials with cement used as a binder. A blade for ceramics, slate, tile and similar abrasive materials has an edge on which tungsten carbide particles have been bonded so as to provide the cutting action, and a knife blade is available for cutting softer materials, including foam plastic, cork, polystyrene, and soft rubber. There is also a range of blades intended to abrade rather than cut, and these are effectively files intended for smoothing and shaping purposes. These are produced with forward edges of half-round, flat and triangular profiles.

Because a jig saw blade is only anchored at one end, its rigidity is limited, and the manufacturers have to arrive at a correct balance between the cross-sectional size on the one hand and its ability to negotiate curves on the other. A wide blade is stronger but less able to saw curves of small radius. In addition, heavier blades can be made longer than the narrower ones, as length affects any tendency to distort. Because of the above considerations, a very narrow blade is produced for where the cutting is particularly intricate involving small radii and sharp bends. These specially narrow blades are around 3mm (⅛in) wide, are of the shorter type at approximately 50mm (2in) working length, and are only really suitable for wood up to about 20mm (¾in) thick. On thicker wood, the cut can easily become out-of-square because of the inevitable lack of stiffness in a narrow blade. This comes about because the user can easily and unwittingly exert side, as well as forward, pressure on the saws as cutting takes place, especially when there is constant change of direction. The thicker the wood, the less easy it is to cut around tight corners, but, fortunately, the need for intricate cutting diminishes as the thickness of the wood increases.

The blade stroke of a jig saw usually measures between 20mm and 26mm ($\frac{3}{4}$in and 1in). For heavy-duty work the longer stroke is preferred. This not only promotes faster cutting, but is also more efficient in the ejection of the sawdust. Manufacturers relate the stroke length and cutting capacity to the power of the motor and overall robustness of the tool. There is a limit to the thickness of wood that can reasonably be sawn with a jig saw. For most practical purposes, this is around 65mm ($2\frac{1}{2}$in), and therefore the variation in the power of the motor is far less in jig saws than with circular saws. Motors for jig saws are rated from around 320 watts to 550 watts input.

138 *Securing the blade on the Hitachi*

The blade is usually locked into the lower end of the main spindle by being clamped by a screw, tightened either by an Allen key or a screwdriver. With one range of jig saws you have to tighten the blade by inserting a long screwdriver through the main spindle, which is hollow. This arrangement promotes accurate align-

137 *Black and Decker model with electronic speed control*

Although there is partial standardization on the shape of the tang of the blade and the corresponding housing on the tool so that the blade can be properly and securely held, more than one system is in use. Therefore, only blades intended for a particular tool must be fitted, as they are not interchangeable from one system of anchoring the blade to another.

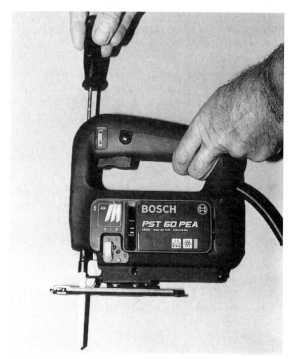

139 *The blade is secured from the top on the Bosch jig saws*

ment of the blade. Whatever method is used, it should always be inserted fully into the spindle so that a positive grip can be made on it. It is also essential to take care that the rear edge of the blade is properly positioned in the guide wheel that helps to control the stroke. The guide wheel is grooved for this purpose, and will benefit from the occasional spot of oil. Because of the flexibility of the blades, they can fairly easily be bent a little out of the vertical. This should be checked before use, and if necessary pressure from the fingers used to restore proper alignment.

Pendulum action

One option now to be found on some models is the pendulum action of the blade, a feature that has advantages and limitations. The amount of 'swing' of the pendulum can be adjusted, usually over a range of three settings, as well as a straight stroke position.

140 *Pendulum selection on the Skil saw*

Even on the maximum setting, the blade moves out of the vertical by only a couple of millimetres or so. The big advantage of this provision is the considerable increase in the rate of cutting. This can rise by two or even three times, and thus counteracts the slowing down effect in thicker timber. The pendulum action allows for a more efficient ejection of the sawdust and

this is especially important where the thickness of the wood exceeds the stroke length of the blade. Where the stroke is less than the thickness of the material, some of the teeth remain within the wood throughout the sawing. Thus the ejection of sawdust is poor. Some gullets are likely to remain filled with dust and thus teeth that do not clear the wood cannot cut efficiently. The effect of pendulum action is to clear the sawdust far more efficiently.

However, if a blade is moving forwards and backwards as the stroke is made, then in effect its width is being increased, and this restricts its manoeuvrability. Maximum pendulum action is ideal for straight cuts or very gentle curves, but not for more intricate work. Thus the setting can be adjusted to suit the curvature of the cut and thickness of the wood, but is not really suitable for complex shapes where the rate of sawing is less important by the very nature of the work, or because the wood is fairly thin and a narrow blade is fitted to the tool.

Scrolling action

Scrolling action is another feature that is offered on some saws. This allows for the blade to be rotated through 360 degress on its vertical axis. Movement is controlled by a knob on the top of the tool that can be rotated while the cutting is taking place; this adds to the flexibility of use. It is a feature that is more advantageous on cutting of a complex nature, but it also aids the ease of use when the saw is operating in a very restricted space. A lever below the knob locks the main spindle so that the blade is in the normal forward position. Some variations also allow for locking at 90 degree steps throughout a full rotation. It is not usual to have a blade guide wheel on a saw with scroll action.

141 *Using the scrolling action*

Speed of stroke

The stroke of the blade can either be at a single rate, or electronically controlled over a range that usually starts at zero. Maximum speeds are around 3,000 strokes per minute or a little more, the figure stated by the makers being the no-load speed, which will drop according to the resistance faced by the blade. For wood, the higher speeds are preferred. Not only does a high rate of stroke provide for a good speed of sawing, smoothness of cut and efficiency are also improved. The slower speeds are to be chosen when materials other than wood are being tackled, although a speed a little less than maximum is advantageous for intricate cutting in fairly thin wood, when controlling the cutting is more important than speed of working.

Guards

Jig saws are relatively safe types of saw, and the user would have to be extremely careless or clumsy to have an accident with one. Nevertheless, some tools, but by no means all, are fitted with a guard to the blade. This can only protect the upper part of the blade and the securing collar. On some tools they are fixed while on others they are rather larger and can either be removed entirely or slid out of the way when the blade is being changed.

142 *Some models have a sliding guard to top of blade*

Sole plates

Almost without exception, the sole plate of a jig saw can be adjusted for bevel sawing. A screw is slackened on the underside, and the plate can then be tilted to left or right, normally up to 45 degrees. On

143 *Adjusting the sole plate on the AEG jig saw*

most jig saws, there is an arrangement within the tilting mechanism that provides for positive location at the normal 90 degree position, and similarly there is an automatic stop at the 45 degree limit of tilt. Some models include a protractor scale for checking the angle of bevel. On others, there is positive engagement at the principal angles.

Bevel sawing

As with the circular saw, there is a loss of cutting depth when sawing is changed from the vertical position to the bevel mode. However, special care is required when bevel sawing, or the angle produced on the wood can be less than the setting of the sole plate. This is because under certain conditions the blade will be influenced by the factor referred to earlier – following the line of least resistance. A blade is cutting more wood when bevel sawing than when cutting the same piece of material square through.

144 *Bevel cutting with the AEG*

However soft the wood and sharp the blade, resistance is offered to the blade as cutting continues. Thus when bevel sawing, the natural tendency of the blade is to flex a little nearer to the vertical, as this means less wood is being sawn. Various factors influence this tendency, including the hardness of the wood, the width and therefore the stiffness of the blade, the sharpness of the teeth, the thickness of the material, and the rate of sawing. In addition, the tendency of the blade to follow the line of least resistance is a little more pronounced when cutting along the grain compared with cutting across the fibres because of the very nature of the grain of wood. Bevel sawing is most successful when a wide, sharp blade is fitted, and when the speed of working is kept fairly slow.

Preventing splintering

Any kind of sawing creates some splintering to a greater or lesser extent on the surface of the wood. This is normally restricted to one side of the material, the surface to which the teeth point. With a jig saw, the cutting action is on the upwards stroke, and therefore it is the upper face where any splintering is going to take place. It is more pronounced when sawing across the grain, and can be a particular problem when sawing plywood and veneered boards. The splitting occurs largely because there is nothing to hold the wood firmly in place alongside the kerf as this is being formed, as inevitably on a jig saw there is a considerable gap between the blade and the sole plate. Because of this, some models of jig saw include a plastic insert that clips into the sole plate and through which the blade penetrates with the minimum of clearance. Thus this insert offers support immediately alongside the blade. Splintering is very much reduced simply because the surface fibres do not have a space in which to lift. The insert cannot be used for bevel sawing, nor is it suitable for pendulum action as the aim is always to keep the aperture as small as possible.

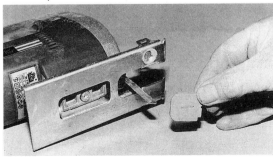

145 *Anti-splinter shoe insert on the Elu*

Dust collection

Jig saws do not create too much dust but, nevertheless, it is always pleasanter to work in as clean an atmosphere as possible. Some models connect to a vacuum dust extractor. This may be via an attachment or by directly linking the hose to an outlet on the body of the saw. As well as providing cleaner working conditions, a dust extractor on a jig saw has a secondary advantage. Because of the upwards cutting action of the blade, dust tends to get deposited on the surface of the wood, obliterating the line being followed. Dust collection helps to minimize this problem.

146 *This Bosch saw has dust collection provision*

Fence

Although probably at its most useful when cutting curves, the jig saw can, of course, be used for straight cuts, and often internal cuts, for which this tool is particularly useful, have straight edges. In order to make actual handling easier, and to improve the accuracy of making straight cuts, jig saws have provision for adding a fence to the sole plate. With

147 *Using the fence on the Hitachi saw*

some models the fence is included as a part of the initial equipment. With others, it is an extra. A screw in the sole plate allows for securing and locking, and on average they will allow for working up to around 150mm (6in) from the edge of the work. Any straight cutting over the limit of the fence can be guided simply by using a straight edge cramped to the work.

Circle cutting

For cutting out circles of wood, or indeed part circles, a circle cutting guide can be obtained for some makes of saw. This slides into the sole plate in a similar way to the fence, and is likewise adjustable. The pin on the attachment is held on the surface of the wood, and thus the sawing has to follow a circular path deter-mined by the setting of the circle guide. Typical cutting capacity is up to around 400mm (16in) diameter, although for repetitive cutting of larger circles a longer guide is not too difficult to improvise.

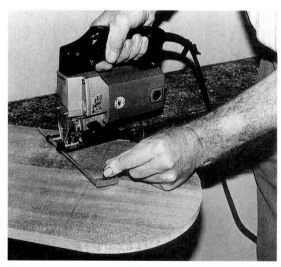

148 *Using the circle cutting attachment with the Makita*

Note that the pin should be positioned in a line square to the blade, and if the inevitable small hole the pin makes is undesirable, then a piece of scrap wood should be secured to the face of the wood on which the pin can locate. The scrap is temporarily secured to the workpiece by using double-sided adhesive tape. This peels off cleanly after use.

Internal cuts

Making a totally internal cut presents a small problem in getting the saw started away from an edge, and there are a couple of ways of overcoming this difficulty. The easier way, where this is practical, is to bore one or more holes in the part that is to be cut

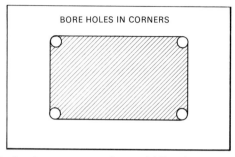

149 *Cutting waste out from middle of workpiece*

away. The holes should be large enough for the blade to pass through when the sawing continues as normal. For a rectangular shape, it is often an advan-tage to bore a hole in each of the four corners of the piece to be removed, and if appropriate, the fence can be brought into use for guiding the saw along the sides of the rectangle.

150 *Making an internal cut after boring holes at the corners*

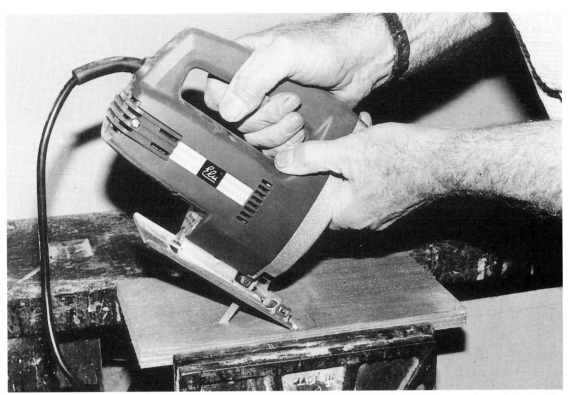

151 *Starting internal cut without holes*

For starting an internal cut without first boring holes, proceed as follows. Rest the saw in a sloping position with the front edge of the sole plate in contact with the wood, and with the teeth of the blade facing the operator. Lower the saw very slowly backwards until the blade contacts the wood. Keep lowering the blade and cut slowly into the face of the wood and

continue all the way through until the sole plate is flat on the face of the wood. This technique works best on wood that is not too thick, and it provides a useful means of cutting into floorboards, and similar *in situ* sawing. Note that the penetrating technique must only be carried out with the blade in the fixed, non-pendulum action, setting. It is dangerous to try it in any other mode.

RELIEF CUTS

152 *Initial cuts made prior to sawing main outline*

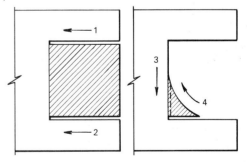

153 *Typical sequence of sawing*

8 | Sanders

Power sanders appear in more types and variations than any other power tool, and there are few sanding operations that cannot be carried out with the aid of power. Hand sanding is a particularly tedious and slow stage of any woodworking project, but power sanders go much further than simply replacing traditional hand sanding. They can carry out shaping operations, levelling off of adjoining surfaces, and virtually eliminate the need for planing during the cleaning up stage.

Materials used

In fact sanding is really a misnomer. Sand as an abrasive grit was only ever used to a very limited extent before crushed glass became the popular material long before the days of power sanding. Glass is cheap and is still used as one of the abrasives for hand sanding, but its limited life makes it unsuitable for use with power. Two materials are now extensively used as abrasive grits. These are aluminous oxide and silicon carbide. Both are products of the electric furnace, and both have a wide range of applications on materials other than wood, including metals, plastics and stone. Sand still lends its name to many things associated with abrading, and words such as 'sanding' and 'sander' remain and are well understood.

The backing material on which the grit is bonded may be either paper or cloth. Cloth is essential for small belt sanders, but paper, which is far more economical, is used on most other types of power sander. The other most important factor concerning abrasives is the size of the particles of grit, as this determines the coarseness of the abrading action and the quality of the surface produced. Related to the coarseness is the speed at which the material is abraded away. This is why sanding is so often carried out in two or more stages, a preliminary stage with a coarse abrasive, and a second stage with a fine abrasive.

Garnet Aluminous Oxide	Emery Cloth			Glasspaper Glasscloth		
UNIVERSAL GRADES	BRITISH	AMERICAN	CONTINENTAL	BRITISH	AMERICAN	CONTINENTAL
	00	000	000	—	—	—
400	—	—	—	—	—	—
320	—	—	—	—	—	—
280	—	—	—	—	—	—
240	—	—	—	—	—	—
220	—	—	—	—	—	—
180	0	000	00	00	000	00
150	FF	00	0	0	00	0
—	—	—	—	F1	—	—
120	F	0	1	1	0	1
100	1	$\frac{1}{2}$	2	$1\frac{1}{2}$	$\frac{1}{2}$	2
80	$1\frac{1}{2}$	1	3	F2	1	3
60	2	$1\frac{1}{2}$	4	M2	$1\frac{1}{2}$	4
50	$2\frac{1}{2}$	2	5	S2	2	5
—	—	—	—	$2\frac{1}{2}$	$2\frac{1}{2}$	6
40	3	$2\frac{1}{2}$	6	3	3	7
36	4	—	—	—	—	—
30	—	—	—	—	—	—
24	—	—	—	—	—	—

154 *Comparative grading chart for abrasives*

Fortunately, we now have the grades of abrasives used with power tools established on a universal basis, and there is close co-operation between the coated abrasive manufacturers and the power tool manufacturers. For most purposes, the grades of 60, 80 and 100, covering coarse, medium and fine respectively, should prove adequate, with grits outside these limits being used for the more special applications. Finer grades, for instance, will probably be required if a lacquered or varnished surface is being flatted down between coats. Where the abrasive has been ready prepared into a belt, drum sleeve, flap wheel, or similar, then just use what is available, as the choice is usually restricted to the three popular grades mentioned.

The belts must be bought to suit the machine. They cannot be prepared on a home-made basis, neither can their sizes be adjusted in any way once manufactured. Although tool manufacturers make their products to suit standard widths of abrasives as produced initially in 50 metre rolls, the lengths can vary from model to model, and a belt of exact length is essential.

For sanders that use abrasive cut from a standard sheet measuring 280mm by 230mm (11in × 9in), there is a wider choice. Use either ready-cut sheets, as offered by the abrasive manufacturers, and by the tool manufacturers who 'buy in' the abrasive products for their sanders, or the full sheet bought and then cut to suit the tool. The latter gives greater scope for using a wider range of grades of abrasives.

The possible exception to the above is on those sanders that use an abrasive sheet specially prepared with perforations. These are usually the orbital sanders, and the perforations minimize the build up of dust between the tool and the workpiece. An excess of dust leads to a loss of efficiency. Orbital sanders that use perforated sheets usually have a dust collecting bag so that the dust can be sucked through the holes and into the bag. However, many of the manufacturers of the sanders that use perforated paper also provide a perforating kit so that standard sheets can be cut to size and then perforated to suit particular tools.

Metallic grits

As well as paper and cloth being used as a backing material, and the two main abrasive grits mentioned, there is another combination available for orbital and disc sanders. The abrasive is tungsten carbide particles, and these are bonded on to a very thin piece of metal. The hardness of these sheets makes them very suitable for a wide range of materials including wood. These sheets are particularly suitable for sanding softwoods, because of the open nature of the particles. This means they are fairly well spaced, which considerably reduces the tendency for the abrasive sheet to clog, which is often a problem with woods of a resinous nature. In addition, the sheets can be easily cleaned by using a wire brush. The sheets are made to correspond to both one-third and one-half sheet sizes, and the discs in diameters of 127mm (5in), 152mm (6in), and 178mm (7in). A variation of the orbital sheets is where the metal which carries the

155 *Cintride's tungsten abrasive sheets*

particles is bonded on to a flexible backing. This allows for easier fixing to the tool, and also provides a slight cushioning effect that can prove useful on uneven surfaces. Tungsten carbide abrasive products are available in coarse, medium and fine.

Belt sanders

The belt sander has become a particularly popular tool. It is very versatile and is extremely efficient. Its abrading action is quite rapid and because of this all belt sanders are fitted with a dust collecting bag, with the option on some models of fitting a suction hose as an alternative to the bag. The size of a belt sander is determined by the width of the belt. These vary between 65mm (2½in) and 100mm (4in), with 75mm (3in) being the most common. Motor sizes vary

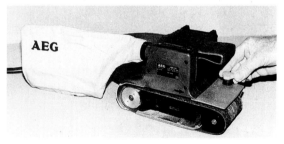

156 *This AEG model has a 100mm (4in) wide belt*

between 550 watts and 1,100 watts. Speed of the sander is measured by the linear amount of belt that passes over the wood, and this varies from around 180 metres per minute, to more than 400. Again, some manufacturers produce models with electronically variable speed. A sander with variable speed is useful on other materials besides wood. Where veneered surfaces are being sanded, and this includes faced plywood and chipboard, great care is required be-

157 *The Black and Decker 'Plus' has electronic speed control*

cause of the thinness of the veneer. This is around 0.5mm ($\frac{1}{64}$in) thick, and can easily be sanded through, especially on narrow surfaces where there is a concentration of effect. As well as using a fine grade of belt – often restricted to grade 100 as has already been explained – a slower speed helps to control the rate at which the surface is abraded.

The belts are held under tension from a spring as they pass over the two rollers, the rear one of which provides the drive from the motor. The standard method of slackening the tension is by pulling out a lever positioned between the two rollers. This moves the forward roller nearer to the rear, and thus allows the belt to be positioned over the rollers, or removed. With the belt over the rollers, the lever is pressed fully home; there is no provision for further adjustment. It is essential to mount the belt so that the arrow on its inner surface corresponds to the arrow on the body, the arrows always showing the direction of rotation. The importance of this is because of the joint on the belt, the lap being arranged so as not to be damaged as the sander is used.

So that the belt runs centrally over the rollers, the

alignment of the front roller must be altered slightly. This causes the belt to move a little to left or right as it is rotating. Too much to the right, as viewed when in use, and the belt will partly run off the rollers, while too much the other way will cause the belt to run against the body of the tool, and the edge of the belt will fray as a result.

Sanding frames

Because of the high rate at which belt sanders work, there is some danger that over-sanding can take place and too much wood is removed. On a large piece of wood in particular, this could result in slight undulations being created that would be difficult to remove even if detected at this stage – often such surface faults only show up when the work has been through the 'finishing' stage of having a surface coating applied. Because of this, some models produced by a small number of manufacturers have provision for adding what is known as a sanding frame, and this may be provided with the sander, or available as an extra.

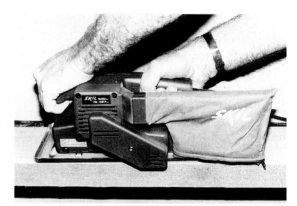

159 *This Skil model comes complete with sanding frame*

158 *Adjusting the tracking on the Bosch PBS 75 sander*

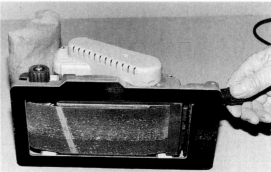

160 *Adjusting the sanding frame on the Elu*

The frame is attached to the underside of the tool and surrounds the belt, the surfaces of both being parallel even as the frame is adjusted. By adjusting the height of the frame relative to the bed of the machine, which provides the sanding area, the projection of the belt beyond the frame can be controlled. Thus the amount of sanding in any one part of the wood can also be controlled and limited in relation to the surrounding area on which the frame rests. The risk of an undulating surface developing when a frame is fitted is therefore virtually eliminated.

Using the sander

Belt sanders need only light pressure when in use. Indeed, excessive pressure can lead to overheating. Generally, sanding is carried out with the grain, and the sander moved fairly slowly over the surface to ensure all parts are treated as evenly as possible consistent with producing a flat surface. There are

161 *Normal working is with the grain*

exceptions to this; adjoining surfaces to the joints of a framed assembly, for instance, are likely to have the grain of the parts at right angles to each other, and therefore constant working with the grain throughout the sanding cannot readily be achieved. The aim should be to carry out the final stages of the sanding in the direction of the grain. In some cases this might be best achieved by means other than the belt sander.

Where a surface requires a considerable amount of sanding, for example when the wood has been sawn on a bandmill but not planed, initial sanding with a coarse grit belt working across the grain can be helpful. This is because cross-grain sanding is quicker than working along the grain, and the belt clogs up less. Initial sanding should be followed with medium and finer grades according to the smoothness required, working along the grain.

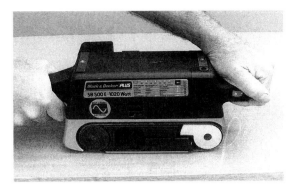

162 *Only gentle pressure is required*

Although cross-graining working is normally quicker, the surface will be left with visible scratch marks. The coarser the abrasive, the more prominent these will be. Working with the grain lessens the scratching effect, and such scratches as are produced blend in with the grain making them far less noticeable. It is largely because of the desire to eliminate scratches, or lessen them as far as possible, that increasingly finer grades of abrasives are used when preparing a surface.

The importance of the direction of sanding in relation to the grain largely depends on what is to happen to the surface afterwards. If the sanding is on a surface that will not eventually be seen, then it matters little. Such an example might be a table top, jointed up from several pieces but requiring some levelling on the underside, as well as, of course, more thorough sanding on the upper surface. For practical and economic reasons, a table top would not necessarily require the same amount of attention on both surfaces, and therefore entirely cross-grain sanding on the underside might be satisfactory. The whole reason for this lies in the type of final 'finish' which is to be applied to the surface by way of polish, lacquer, varnish or similar coating. Clear, or transparent, surface finishes do not just show the wood beneath, but tend to highlight any faults in the wood. Thus any scratches from cross-grain working would be emphasized, and could also lead to problems of colour matching, especially if the wood was stained. A scratch across the grain disturbs the surface fibres, which as a result become more absorbent, and the more stain or other finish, even if clear, is absorbed, the darker it will become. Preparation for clear finishes must be thorough, final sanding always along the grain as far as this is practicable and with a fine grade of abrasive. Whatever finish is used, this cannot make up for poor preparation.

The position is rather different where the surfaces being levelled and smoothed are to be given an opaque finish such as paint. Because of its opaque nature, paint obliterates the surface and therefore, assuming proper painting procedures are being followed, surface scratches will be concealed. Excessive and deep scratching, and minimal painting, can still result in the scratches being visible, and proper preparation relative to the nature of the finish must be followed. On the other hand, too smooth a surface for painting is not always a good thing; some scratches help to provide a key for the paint, and thus improve the bond between paint and wood, reducing the possibility of subsequent flaking.

Stationary use

For some sanding operations, and especially where the workpieces are small, it is easier to have the sander stationary and move the wood against the belt.

163 *The AEG ready for inverted use*

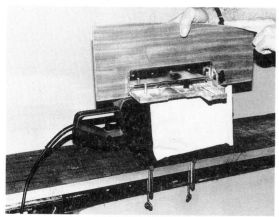

164 *The fence keeps the wood vertical*

Because of this, certain models of belt sander can be held either inverted or in other positions to allow for this mode of working. It is necessary to have the appropriate stand for the sander, the stand usually

165 *Fence set across the wood for small items*

allowing for the tool to be either horizontal or vertical. For stationary use, it is also necessary to have a fence attachment for the sander. This can normally be secured either parallel to the belt, or at right angles to it, and the fence can be adjusted from 90 degrees to 45 degrees. Although stationary use is particularly suitable for sanding small pieces of wood, it can also be used for larger material and with the belt in the vertical position provides a useful means of trimming the curved edges of workpieces. While the fence is associated with stationary use, this attachment does have a purpose when the work is fixed and the sander is being moved, especially on fairly narrow edges. Balancing the sander on a narrow surface is never too easy, and the fence therefore ensures that the edges are being maintained at the correct angle to the face during sanding, whatever this angle is.

166 *The Bosch allows for vertical mounting*

General

What has been said so far about the direction of working in relation to the grain, and the importance of this as a proper preparation according to the nature of any subsequent finish that is to be applied, does not just apply to belt sanders. Many other types of sander are less 'directional' in their operation, but the question of the grain of the wood and the need to avoid scratches when a clear finish is to be applied, must be kept in mind with any sanding process.

167 *Wire brushing helps to clear clogged surface of belt*

Brief mention has already been made regarding cleaning abrasive materials with a wire brush. The abrading action of the belt or sheet is considerably reduced if the surface becomes clogged with particles of wood, and therefore removing these can considerably extend their life , but never attempt this when the tool is running. Brushing in various directions is the most effective method. An alternative to brushing to restore the abrading surface is to hold a piece of plastic to the belt or sheet, this time with the sander actually running. The plastic should be of the semi-hard but flexible type, and plastic hose pipe has been found to be ideal. This has the property of lifting the wood debris which is trapped by the particles of abrading grit.

Orbital sanders

Orbital sanders are often called finishing sanders, as they are best suited for lighter sanding operations rather than the heavier work that the belt sanders can tackle. They are in fact ideal for the secondary stages of preparing a surface, but are less suitable for

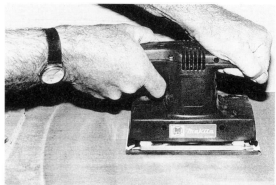

168 *The Makita orbital sander*

levelling purposes where, for example, adjoining surfaces of an assembly are not in alignment. Although with the appropriate abrasive sheet in position they are capable of producing particularly smooth surfaces, their rate of wood removal is quite low. Largely because of this, the need for a dust bag is less critical and provision for dust collection is not therefore provided on many models.

169 *Black and Decker sander with variable speed*

The orbit action is in fact a circular movement, the diameter of the orbit ranging from around 1.5mm to 6mm ($\frac{1}{16}$in to $\frac{1}{4}$in). Speeds of orbit are up to 10,000, but these sanders do not absorb a lot of power and motor ratings are usually between 135 watts and 350 watts. A few orbital sanders have variable speed control, a useful refinement for anyone who does a lot of work of a particularly delicate nature. Those models that do have provision for dust collection normally have holes in the sanding platen to allow for the extraction of dust, and therefore use sheets with matching perforations. A few sanders that have dust collection also provide a frame that surrounds the sanding sheet. This is totally different from the sanding frames referred to for belt sanders. They do not control the depth of sanding but merely act as hoods to aid the collection

170 *Some models use perforated paper to aid dust collection*

171 *This Makita model has removable dust retaining frame*

172 *Orbital sanders are best when used for finishing purposes*

Once again, only light pressure is needed, with all parts of the surface being treated as evenly as possible consistent with the degree of smoothness required. Extra care is called for on narrow edges because of the possibility of tilting the tool, and while both hands are normally used to operate the sander, single-handed use is possible in most instances. The orbital sander does not lend itself to use in the stationary mode.

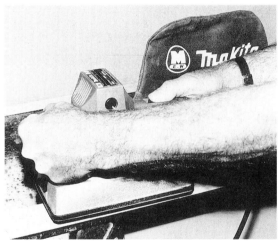

173 *The narrower the surface the lighter the pressure*

of the dust. The frame does impose a slight restriction in the operation of the sander, but their use in any case is optional.

Orbital sanders are classed as being either one-third or one-half sheet size, meaning the size of abrading paper used is this fraction of a standard 280mm by 230mm (11in × 9in) sheet. Methods of securing the paper, or other abrading sheet, to the tool vary. Most depend on a spring clip at each end, and changing sheets is quick and easy. It is, however, essential to ensure that the ends of the paper or other material are properly anchored in the clips, and in such a way that it is reasonably taut against the platen. Care pays dividends here, as improper mounting will result in creasing and tearing of the paper, and an unsatisfactory sanding action.

The circular action of this sander means that for each orbit, sanding is taking place at every angle to the grain through 360 degrees. Thus there is considerable cross-grain working with possible scratching. In fact, the high speed of the orbiting of the sander, combined with the movement of the tool across the workpiece, leaves the surface remarkably smooth. Such scratches as are produced are so fine as to be insignificant.

Palm sanders

A variation of the orbital sander is the palm sander. These are really small versions of the one-third and half-sheet models, and are specifically intended for single-handed use. They take a much smaller sheet of abrading paper, also held by spring clips, and may be

174 *The Black and Decker palm sander*

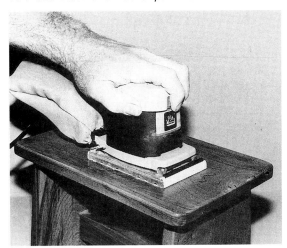

175 *Sanding during polishing, this Elu model has dust bag*

176 *Perforating kit supplied with certain sanders*

obtained with or without a dust bag. Those with dust collection have a perforated base and use appropriate paper. The palm sander is especially useful for small and delicate items, veneered surfaces, marquetry, and similar decorative work. This sander is also ideal for the flatting down stages necessary between successive applications of finishing coatings, including the final flatting of catalyst-type lacquers before the burnishing stage. Correct grades of abrasive are important at all times, and especially when the sander is being used during the finishing process.

Disc sanders

One of the simplest and cheapest sanders for surface preparation is the disc sander. In its early days this was often known as the flexible disc sander, as the backing to the abrasive is a thick rubber pad with a metal fitting in the centre that connects directly into the chuck of the power drill. A drill with a side handle is essential, and one with dual or variable speed is desirable. The abrasive is normally secured by a single screw with washer, which is tightened into the metal compoment.

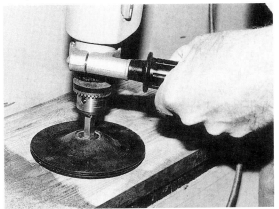

177 *Rubber-backed disc sander in use*

Disc sanders of this type can remove wood quite fast, which can prove to be both an advantage and a drawback. It is very easy to over-sand and create curved depressions in the wood. Indeed, it requires some practice and skill to use a disc sander to produce a surface that is even reasonably flat. Circular score marks are usually left on the surface with this type of sander unless a lot of work is done with finer grades to remove them. This sander is useful for preparing softwood items for painting providing the quality is not too important, and is also useful for sanding down old painted surfaces, especially where the paint is very hard.

178 *Cintride all-metal sanding disc*

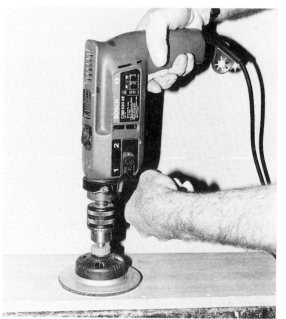

180 *The special collar allows the disc to 'float'*

179 *The Wolfcraft lock on-off sanding disc*

Because of the way the disc is connected directly to the drill, the sanding action is very largely controlled by the movement of the drill and especially the angle at which this is held. The underside of the rubber pad tends to become slightly convex in use, and it is normally used at an angle so that only a part of the abrasive is in contact with the wood. Too great an angle can lead to ridges being formed.

A much improved pattern of disc sander is the type that incorporates a flexible knuckle between the pad and the drill. A screw on this collar allows for the pad to be either rigidly locked in relation to the drill chuck, or to have a floating action of up to around 20 degrees from the axis of rotation. Furthermore, the main part of the disc is rigid but with a foam cushion on the underside to which the abrasive paper is secured. When the nut is set to allow floating action, the risk of the outer part of the disc removing excessive amounts of wood and thus causing deep scars is eliminated. In addition, the whole of the surface of the abrasive disc can be brought into contact with the wood. This gives increased effectiveness and means that the sanding action is far more uniform. With finer grades of paper, the scratching effect across the grain is reduced very considerably and high levels of smoothness can be achieved.

181 *The disc can also be used for sharpening purposes*

Orbital disc sanders

A different type of disc sander is the pattern known as a random orbit sander. This is not an attachment but is a single purpose tool. As well as the 150mm (6in) disc rotating, it also follows an orbital path that has a diameter of approximately 10mm (⅜in). The disc is made of plastic, which flexes to a limited extent, and has a cushioned face. The abrasive discs are of the self-adhesive type.

183 *Mini sanding discs with and without foam backing*

182 *The Black and Decker professional random orbit sander*

The advantage of the random orbital sander is in the quality of the finish, combined with a good speed of abrading action. Because of the dual movement of this tool, the scratching effect of the standard pattern of disc sander does not take place, or at least it is reduced to a negligible amount. The fact that the disc does not follow a circular path means that the risk of the periphery of the abrasive creating curved depressions is also eliminated.

Sanding pads

Mini versions of the disc sander are also produced, and are particularly useful to the woodturner and the carver. They are produced in diameters of 25mm (1in), 50mm (2in) and 75mm (3in), and in two variations. One has a soft foam pad around 19mm (¾in) thick onto which the abrasive is attached, the foam being bonded to a hard rubber base into which is moulded a steel shank. The alternative pattern simply does not have the foam.

The foam cushion type is especially useful for sanding hollow surfaces, as the foam largely accommodates itself to the concave shape. If you are making bowls you will find these especially helpful at the sanding stage. They are used with the lathe running, and the combination on the rotation of both the wood and the abrasive quickly produces an excellent surface. Normal sanding by handholding the abrasive paper against the rotating bowl can cause scratching and is often an awkward job depending on the shape of the work. Using a mini disc of appropriate size and grade can solve many of the problems of obtaining a good surface on bowls and similar hollow faceplate turning. They are equally useful to the carver, especially when the carvings are of the sculpture style.

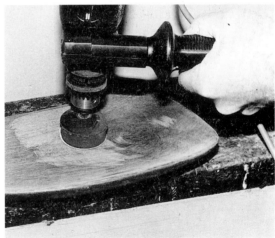

184 *Using foam-backed mini disc on carved dish*

Abrasive discs as fine as 400 grade can be obtained for all three sizes of disc. These sanders can be used mounted directly into the drill, but for many applications their use in a flexible drive shaft will be preferred. The abrasive discs are secured by Velcro.

Stationary disc sanders

The stationary disc sander has already been mentioned in an earlier chapter as a piece of home-made equipment. This pattern of sander can be obtained specially for drill-operated use, a diameter of 175mm (7in) being typical. Designed for cramping to a worktop, the table will tilt up to 45 degrees from the horizontal. A fence is provided, which is also adjustable and can be locked into a stationary position, or set in such a way that while the angle to the disc is maintained, it will slide across the table.

185 *Static disc sanding attachment from Wolfcraft*

This type of sander is ideal for trimming purposes, and especially on end grain surfaces, which are difficult to tackle by other means. Convex edges, after being sawn with a jig saw, for instance, can be brought to both a smooth surface and uniform shape on this equipment, and mitres can be adjusted to improve the fit. Toymakers, and others whose work involves small pieces of wood, will find this pattern of sander especially helpful, and the tool can also be used for trimming the outer surfaces of small, box-like assemblies.

As far as possible, only the half of the disc that is moving downwards should be used so that the rotation of the disc holds the wood against the table. Great care is needed whenever the whole of the abrasive area is used because of the risk of the work being lifted from the table. In addition, wear on the paper is far less if the wood is kept slowly on the move as abrading takes place.

Sanding drums

Sanding concave edges cannot readily be achieved by sanders designed primarily for working on flat

186 *Small sanding bobbins*

surfaces, and the sanding drum provides one such method of tackling curved work. Drum sanders are around 75mm long by 63mm diameter (3in × 2½in), with a central shank and nut providing the means of mounting the drum and expanding the plastic cylinder so as to grip the abrasive sleeve. Similar but rather smaller versions are usually known as bobbin sanders.

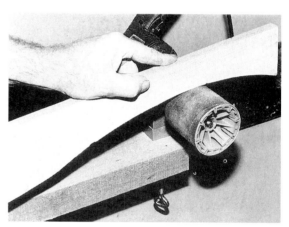

187 *Sanding drum of 75mm (3in) diameter*

One way of using the drum sander that is powered by a drill is when the latter is secured to the workbench by the use of a bench stand. The work is then passed over the drum, but the direction of working is important. This must be against the direction of rotation, or control will be difficult. An alternative way of using both drum and bobbin sanders is when the drill is in a drill stand. A false table of wood should be added to the base of the stand for this, with a hole first made in the false table. The hole should be rather larger than the drum or bobbin, so that the abrading surface can be lowered slightly below the top of the table. This gives proper support to the workpiece, and ensures that the abrading sleeve is in full contact with the surface of the wood requiring smoothing. This method also allows for adjusting the height of the

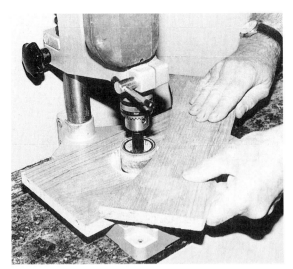

188 *Using sanding bobbin with drill on stand*

drum so that wear is not just concentrated in one part of it.

A variation of the drum sander is the type composed almost entirely from a fairly dense foam plastic that provides a flexible cylinder on which the abrasive sleeve is mounted. It is held in place simply by pressure from the foam, and this attachment is ideal for shaped work such as the hollow in the seat of a Windsor-type chair. It can also be used in a drill stand in a similar way to that described above.

189 *Flexible sanding drum*

190 *Freehand use of flexible sanding drum*

Flap wheels

Abrasive flap wheels are particularly useful on uneven surfaces, as they can adapt to irregular contours to some extent. These too are available in large and small sizes, but it is the tips of the multiple abrasive flaps that do the abrading, and as they wear, the diameter is simply reduced so that they remain effective until

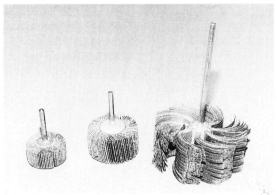

191 *Flap wheels of various sizes*

worn well down in size. One pattern of flap wheel has the abrasive flaps cut into narrow strips. This makes them particularly useful for sanding mouldings and similar profiled surfaces.

192 *Serrated flap wheel in use*

Sanding files

Sanding files also depend on a sanding belt but are quite different from the belt sanders already described. Widths of belt vary from around 12mm to 28mm ($\frac{1}{2}$in to $1\frac{1}{8}$in), the belt running over two wheels with a small pad between. Tension adjustments of the belt and belt replacement are, of course, readily made.

These sanders provide a means of working at close quarters and in tight spots difficult to tackle by other methods. You can sand small internal cut-outs with them, and the narrow belt model provides an easy and effective way of cleaning out the mortise as required for a mortise lock. With a power file mounted in a stand, a means is provided of freehand sanding irregular shaped work by manipulating the wood across the belt.

193 *The Makita sanding file*

195 *Sanding file in bench-mounted mode*

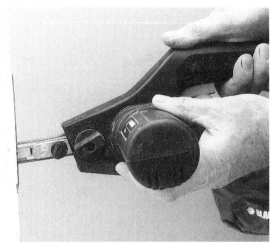

194 *The Black and Decker sanding file trimming lock mortise*

9 | Planes

As power planes have improved in their design, and with refinements constantly being introduced to add to their versatility and ease of use, the popularity of these tools has grown considerably. Not only do they have a huge appetite for work, executed with little physical effort, all but the smallest and simplest of models will carry out rebating. They are, though, only really suitable to use when planing surfaces that are narrower than the width of the cutters. On wide boards, therefore, although the surface can be planed with a portable plane, the degree of accuracy that can be reached is lessened considerably. By their length as well as their width, power planes are at their best on smaller pieces of wood, and they cannot be regarded as alternatives for static surface planing machines, or plane-thicknessers, because of the width of the cutters on these machines, and particularly the length of the infeed and outfeed tables. Length of table on a machine, or sole on a portable plane, is a key factor in their ability to true up longer pieces of wood.

Plane sizes

The size of a portable plane is determined by the widths of the cutters or blades, the most popular size being 82mm (3¼in), which provides comfortable working on wood up to 76mm (3in) in width. Lengths average around 305mm (12in). Some models are shorter while one or two long versions are available up to 450mm (17¾in). A small number are produced with larger cutters, up to 150mm (6in). The next critical factor is the depth of cut, that is, the amount of wood that can be removed in one 'pass'. This varies from 0.5mm to 3.5mm ($\frac{1}{64}$in to $\frac{5}{32}$in), so clearly the power of the motor has to be related not just to the width of the cutter, but to the depth of cut. However, while the plane with a small depth of cut can be expected to plane at its maximum width and depth, it is rare to want to plane 3.5mm ($\frac{1}{8}$in) deep over the full width. Deep cuts are almost always likely to be confined to narrow edges of boards less than the width of the cutters. Where a wide and very deep cut is wanted, speed of handling can be slowed down in order to maintain full control. Because of this, the power of motors varies from around 350 watts to 1,100 watts. Allied to both depth of cut and motor wattage is the speed of rotation, which can vary from 12,000 rpm to 20,000 rpm. As there is universal adoption of cutter blocks holding two blades, the number of cuts per minute is double the rpm. There is some loss of speed under load, but even so, portable planes rotate at around three times the speed of a planing machine.

The importance of the rpm, or more particularly the number of cuts per minute, lies in the quality of the finish produced. Because the cutters follow a curved path as they plane the wood, the surface produced is not totally flat but consists of a series of scallops. The greater the number of cuts per minute, combined with a moderate rate of feed, the closer and smaller the ripples, and with properly set and correctly used planers, the ripples become so fine as to be undetectable, and normal sanding procedures remove any traces. Ripples are more likely to be visible on the slower revving planing machines, especially when the rate of feed is high.

Depth of cut

The depth of cut is governed by the relative height of the front and rear parts of the sole, or more correctly by the height of the front of the sole in relation to the cutters. Both cutters must have exactly the same projection from the block so that their cutting edges are following precisely the same path, with the rear part of the sole exactly level, or tangential to, the cutting edges. If, for any reason, they project too far, planing will be unsatisfactory and the surface poor, resulting in a 'step' at the end of the cut. If the cutting edges are below the sole, then the planing action will tail off soon after the start so that nothing is being removed. The adjustment of the height of the front part of the sole is governed by rotating the knob at the

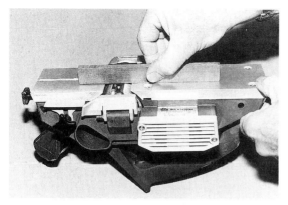

196 *Checking blade and table alignment*

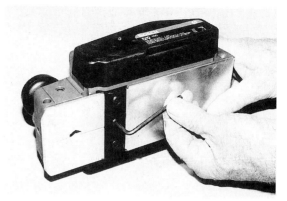

198 *Tightening blades on Bosch Plane*

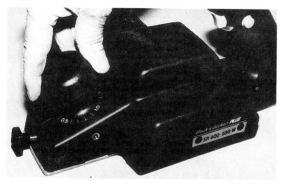

197 *Depth of cut control on the Black and Decker 'Plus' plane*

final tightening will alter the setting. The blades must always be checked against a wooden straightedge, placed on the rear part of the sole and with the plane inverted. The straightedge should overlap the cutter block, so that as this is rotated by hand, the cutters will touch, or miss, the straightedge. The cutters should just touch the wood, but only just. If they move the straightedge forward by 2 or 3 mm ($\frac{3}{32}$ or $\frac{1}{8}$in), then this is acceptable, but a greater amount of forward movement means they are projecting too far. Both ends of both cutters must be checked. In fact cutters with a built-in system of location will normally position themselves without trouble, but not all cutters are of this type. If ever the cutting action of a planer is unsatisfactory, the setting of the cutters should be checked as this is the key factor in the efficient cutting of a power plane.

forward end of the body of the plane, this knob also acting as the second handle. There are usually some markings on the knob whereby the adjustment to the cutting depth can be checked. On others, the knob is rotated via a series of notches, with each notch representing a given amount of movement of the sole. What really matters when any kind of surface planing is taking place is not the actual measured setting of the tool, and therefore the amount of wood being removed, but whether the setting is appropriate to the work in hand.

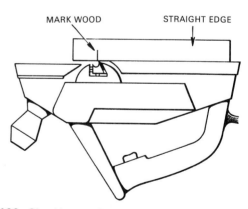

199 *Checking setting of blades*

Blades

Most portable planers have blades with a groove on the back, which provides for positive location within the block. Rather different systems are used to lock the blade to the block, with tightening being either by spanner or Allen key. Provision is made by way of a couple of jacking screws to allow the projection of the blade to be controlled; adjustment of these along with tightening of the blades must be made together, or

Blades that are of the above pattern are normally reversible, and are of the disposable type. They are honed as a part of the manufacturing process, with both edges being available for use. In addition, TCT blades are now the rule rather than the exception, and these cannot be sharpened by normal means. In any case, repeated honing of a blade with built-in location provision would make accurate setting impossible. TCT blades have a long life for most uses, but there are exceptions even for these blades. Those timbers that are particularly gritty can have a severe blunting effect on a plane blade of any type. Teak is notorious for this, as it has a high silica content in the pores of the grain, which has a marked abrading effect on plane and similar blades and cutters. The blunting effect of abrasive timbers is far less on circular saw blades, which in any case can be sharpened more readily even though specialist equipment is required for TCT blades. The other danger to plane blades is the possibility of foreign bodies in the wood – nails and screws in particular. The price that is paid for most extremely hard materials, including tungsten carbide, is that they are also brittle. This applies especially to plane blades where the cutting edge is naturally quite thin. If a metal object is struck, there is a very high probability that the edges concerned will be rendered useless for further work. As far as possible, using a portable power plane on a painted surface should be avoided. Not only does the paint itself have a noticeable blunting effect, it also conceals what could be potential dangers for the blades. Where used material has to be planed by a power plane, close examination should be made for nails, screws and fittings.

Blades that do not have locating grooves are also likely to be of high speed steel, and non-reversible. They are, therefore, intended to be resharpened on blunting, and must be set in the block and carefully checked for alignment as described. Although HSS will blunt quicker than TCT blades, they are less likely to suffer severe damage from hitting metal objects, and damage can be ground and honed away.

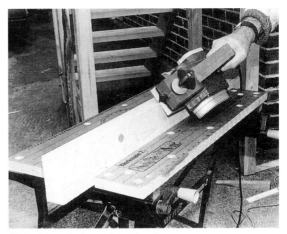

201 *Chamfering with Elu plane*

different sizes for this purpose. The reason for this is that the grooves influence the size of chamfer that can be produced, as a second pass cannot be made using the grooves as a guide. It is, of course, possible to make chamfers of all reasonable sizes without the use of the grooves.

Chipping collection

Power planes are capable of producing a lot of chippings in a short space of time, and because of this, most models provide for connection to a vacuum hose. In most cases, though, it is necessary to obtain the appropriate adaptor. Chipping collection bags are very much the exception with planes. They have limited use because of their inevitable small capacity compared with the waste created. It is a consideration for the person who wants a planer for on-site work where debris must be kept down to a minimum.

Chamfering facility

A feature on many planes is a V groove on the front sole, which is used when forming chamfers. Indeed, at least one manufacturer's planes have two grooves of

200 *Two chamfering grooves on this Skil planer*

202 *The Skil has chippings' bag*

Using the plane

For normal planing, it is essential that at the start of the cut the pressure is placed at the front of the tool. The front part of the sole must be in close and full contact with the wood, with the planer being held horizontal.

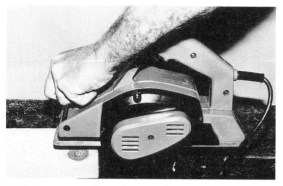

203 *Pressure at front is essential at start of stroke*

Without deliberate pressure on the front, the tendency is for the weight of the plane to cause the rear to drop slightly. This would result in the plane pointing slightly upwards at the start of the planing action, then being corrected by the plane itself as the rear of the sole becomes supported on the wood as it is moved forwards. Only a slight gap beneath the front of the plane is sufficient to cause problems. Lack of pressure on the front of the plane causes excessive planing at the start of the cut. This results in a rounding over at the end of the wood.

At the end of the cut, pressure is required at the rear of the plane, to ensure that the sole plate remains in contact with the wood. Because of the distribution of the weight of the plane, and the greater length of the rear sole plate, this is fairly easy to achieve, although a lack of awareness or concentration can result in the plane tipping forward slightly at the end of the cut. This is very much the same as for using a hand plane,

204 *The Elu has a 3.5mm (⅛in) depth of cut*

where the downwards pressure is transferred from the front to the rear during the course of the stroke. Failure to apply pressure at the right place at the right time is sure to result in the surface becoming rounded, whatever type of plane is being used. It is easier to make the error of producing a convex edge than a concave one. The plane will cease to cut if the latter is attempted. Whatever the depth capacity of the plane, it is wise to have it set fairly fine, say around 0.5mm ($\frac{1}{50}$in), when a piece of wood is being initially trued up. Two light cuts are usually better than one heavy one. The aim when trueing up is to produce as flat a surface as possible with the minimum amount of wood removed.

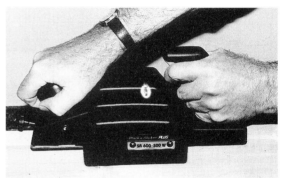

205 *The Black and Decker planer in use*

A powered plane will not produce a true surface automatically. This depends more on the operator than the tool. If a surface is bowed at the outset, simply to plane the wood from end to end will make it smooth and reduce it in size, but not make it flat. It is always wise when trueing up a piece of wood to look along it from end to end before starting the actual planing. Any distortions in the wood can usually be spotted by eye, and the planing tackled accordingly. Generally, it helps if the hollow surface is planed first, as the plane, if properly held, will have a natural tendency to remove more wood from the high parts, and less from the hollow or low part. Checking is just as important as the actual planing. This should be done by eye, and with a straightedge where this is appropriate. An excellent way of testing how flat and straight a surface is, is to place two similarly planed surfaces together. Any gap between the two surfaces means that one or both must be inaccurate.

If, for any reason it is desirable to start trueing up a distorted piece of wood by planing the convex surface, then the basic principle of procedure is always the same, that is, concentrate on the high parts, and avoid the low. Thus for a rounded surface, planing should commence in the centre, with each stroke becoming progressively longer with the final skimming being from end to end. When a high standard of accuracy is wanted, then testing should be carried out as the planing proceeds, not simply left until the end.

Fence

When it is desired to follow the normal sequence of preparing a piece of wood from being rough sawn to planed, then the first surface to be planed is one of the sides and always referred to as the face side. This is followed by one of the edges, known as the face edge.

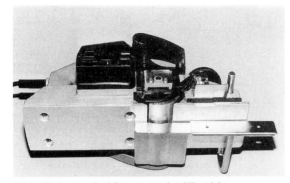

206 *Right-angled fence on the Hitachi*

As well as being straight, the two surfaces must be at right angles to one another. Most planes are now supplied with a fence. This is set at 90 degrees to the sole or can be adjusted to this angle. The face can also be located at any position across the width of the sole. Using the fence when properly set and adjusted means that the second surface will be planed square to the first, providing the fence is kept in close contact with the face side. Obviously, the same use of the fence is adopted when it is just the edges of a wider, ready-planed board that are being tackled by the power plane.

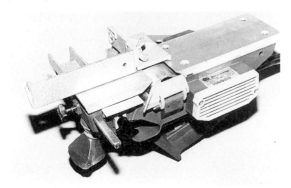

207 *Adjustable fence on the Elu*

Rebating

One operation that the portable plane is ideal for is rebating. All these planes have cutters extending the full width of the sole, and are normally provided with fences, so they can form rebates according to their own limits. While one restriction to the size is the extent of the blades, the other is the depth, and this varies from model to model. Generally, those planes with less powerful motors and limited depth of cut also have smaller capacities for cutting rebates, and conversely the rather better quality planes can cut deeper rebates. The maximum depth of rebates that can be cut varies from around 8mm to 25mm ($\frac{5}{16}$in to 1in).

208 *Bevelling with the Bosch with fence in position*

When cutting a rebate, the fence must be set so that the amount of the cutters left exposed equals the width of rebate required. The depth of the rebate, though, has to be judged by the number of passes made combined with the depth of cut as set by the adjusting knob. The depth of the rebate, therefore, is best marked on the wood and an ordinary marking gauge is ideal for this. Thus planing continues until the line is reached. For repetition work, counting the number of passes can be useful.

It is easy when forming rebates to produce one where the depth is not the same at both ends, but tapers downwards towards the far end of the wood. This is caused by the planing action not starting right at the near end of the wood, but slightly in. Thus the near end of the wood simply has less material removed. This is one reason why the gauge line is

209 *The Hitachi in the inverted mode*

useful, especially until the technique has been mastered. If the vertical surface of the rebate, as this is being progressively cut, develops a series of steps resulting in a sloping edge, then the fault is almost certain to lie in the setting of the cutters. It is essential for cutting clean rebates that the ends of the cutters be accurately set in relation to the right-hand side of the body. The ends of the cutters must be level with or fractionally beyond the body. This in no way affects the setting of the cutters already described. It is solely a lateral adjustment. The steps in the rebate are a result of the blades being set with their ends within the right-hand edge of the body.

It has already been said that the depth of the rebate has to be judged and controlled by the user. There is in fact the odd exception to this as a very small number of planes are equipped with an adjustable depth stop. This is fitted to the right-hand side of the body, and when this reaches the surface of the wood, then planing ceases. However, some care is needed because as the stop is on the extreme right of the body, it

210 *Adjustable rebating depth stop*

is possible for the plane to become tilted a little to the left, and for planing to continue resulting in a sloping surface to the rebate.

Usually when rebating, the wood is held in the vice or something similar, such as a Workmate. It is possible that as cutting the rebate progresses, the fence of the plane fouls whatever is holding the wood. This can result in the depth of the rebate being erratic, and the amount of wood projecting above the top of the vice should be checked at the start. Small pieces of wood are best rebated as described below. In fact, whenever a power plane is being used, the wood must not simply be properly supported, but positively held as well. This is because the rotation of the cutters on the wood can cause it to move forwards. Indeed, it is certain to do so if not restrained.

Stationary use

Many manufacturers offer a stand whereby the planer can be mounted in the inverted, fixed position, thus enabling it to be used as a small surface plane. The guide fence is essential when the plane is used in this way, so that material can be faced and edged with an accurate right angle between the two. While the

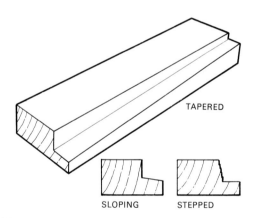

211 *Common faults when rebating*

remaining two surfaces could simply be planed in a similar way, this would not necessarily bring the wood to the correct width and thickness. For accurate dimensional preparation of the material, as well as producing flat surfaces, it is best to combine this with the use of a circular saw. The procedure is then to plane the face side and face edge on the power plane, saw to width and thickness but slightly oversize, then plane this side and edge to produce smooth surfaces all round and accurately to size.

On a planing machine, the fence is on the right-

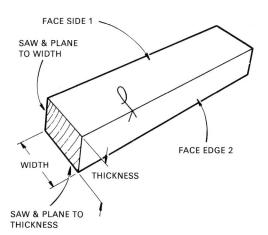

FACE SIDE 1

SAW & PLANE
TO WIDTH

FACE EDGE 2

WIDTH

THICKNESS

SAW & PLANE TO
THICKNESS

212 *Trueing up wood on all surfaces*

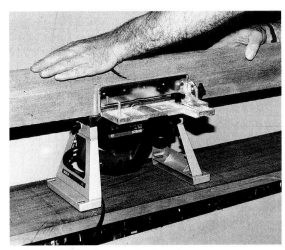

213 *Static planing on the Bosch*

hand side of the tables, whereas on an inverted power plane, it is on the left. In practice this makes no difference. The first surface to be planed must be kept tight to this when the edge is being prepared. Because of the small size of the front part of the sole, it is essential that full contact be made on this with the wood before moving it over the cutters. Planers intended for inverted use have a spring-loaded guard that automatically covers the cutters except when a piece of wood is passing over them. The guard is mounted either on the plane, or the stand.

The guard is provided for obvious reasons, and a power plane used on a stand should not be operated without one. Even so, the fingers must be kept well clear of the cutters at all times, and a push stick used as the end of the wood is passing over the danger area.

In the inverted mode, the plane is particularly useful for preparing small pieces. These are awkward to plane when on the bench because of the difficulty of holding them. Conversely, it is not satisfactory to plane long pieces in this way, because of the difficulty of balancing a weight of timber on a small surface, especially at the start and completion of an operation. The plane when inverted also provides a safe and satisfactory way of rebating small pieces that are otherwise difficult to tackle. Apart from the obvious reversal of what moves and what is stationary, what has been said already about cutting rebates still applies when rebating is carried out in this way.

Ever resourceful, some of the manufacturers produce alternative sets of blades that are wavy along their cutting edges. The idea is that they will produce a textured, rustic surface, which, while not reproducing an adzed effect, do provide an alternative to the normal, totally flat surface.

10 || **Routers**

The router is probably more versatile than any other power tool, and can carry out a wide range of operations and functions, though it is true that much of its adaptability depends on the appropriate cutters being available. Cutter design and technology are constantly being developed to widen further the usefulness of this tool. Sundry equipment for use with the router has also advanced a lot over recent years, some of this of a specialist nature, such as the stair string routing jig, others, including the tracking fence, suitable for more general use. The router is the kind of tool whose usefulness is also partly dependent on the ingenuity of the user, as much work can be achieved by the use of home-made jigs and templates. And, of course, the router itself has undergone many improvements since it was first introduced many decades ago as an industrial tool but with limited application, and refinements continue to be developed and incorporated.

Router speeds and capacities

The router is a high-speed tool. No-load speeds range from 18,000 rpm to 28,000 rpm. The high speed is necessary in order for the cutter, or bit, to cut the wood in a clean way leaving the surface relatively smooth, but more importantly, free of torn grain. The high revving of the tool is also essential because the cutters can be of small diameter, as little as 1.5mm ($\frac{1}{16}$in), and with a single cutting edge. Many of the more complex cutters have diameters of 50mm (2in) and above, with multiple cutting edges. As is explained in Chapter 6 (p. 62), the diameter of the cutter or blade and the rpm of any tool are closely related, as it is the speed of the cutting edge or edges as they pass through the wood that is the critical factor. Because cutters now tend to be developed with larger diameters, there has become a corresponding need to provide routers with slower speeds. The answer to the problem of retaining a router that will handle cutters over a wide range of diameters and with equal efficiency is to provide variable speed, and some routers now have this as an

214 *Hitachi heavy-duty router*

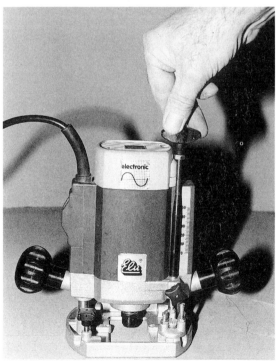

215 *The Elu MOF 96 electronic model*

electronically controlled feature, starting usually with a lower speed of around 8,000 rpm. The variable speed option has the added advantage of allowing for a 'soft start', gives better control when the router is being used freehand, and adds to its range of usefulness when materials other than wood are being tackled.

The larger the cutter, the greater the amount of power that is required in order for it to perform properly, and without too much loss of speed. Power input on routers varies from around 400 watts to 2,000 watts and even higher on the heaviest industrial tools. Clearly, there should always be adequate power to suit the operation being carried out, and to some extent the makers influence this by the size of chuck fitted to the router. This is of the collet type, the three common capacities being the imperial sizes of $\frac{1}{4}$in, $\frac{3}{8}$in and $\frac{1}{2}$in. The lower powered tools have $\frac{1}{4}$in chucks, the heavy-duty ones being fitted with $\frac{1}{2}$in chucks, and some from the middle of the range having $\frac{3}{8}$in capacity collets. It must be emphasized that the $\frac{3}{8}$in and $\frac{1}{2}$in capacity chucks are the maximum, as collet cones of $\frac{1}{4}$in and $\frac{3}{8}$in are available enabling larger chucks to be adapted.

Collets

The collet size must correspond to the shank of the cutter. There is a very wide range of cutters produced with an appropriate size of shank, the number of cutters being rather less as the shank size increases. Shank diameter is very much related to cutter diameter in order to provide not just the strength, but rigidity and therefore freedom from vibration. Although the majority of cutters are made to the imperial sizes stated, some are produced to metric diameters. These must be used with collets of corresponding size, as interchangeability between imperial and metric shanks and collets is not possible.

Most collets require a couple of spanners to tighten and loosen the collet, one to lock the spindle and the other to turn the chuck ring. It is essential to insert as much of the shank of the cutter as possible in order for a proper grip to be made. Trying to extend the reach of the cutter by only having the shank partially inserted is highly dangerous because of the risk of the cutter working loose and even flying out. Note that on some routers there is a deliberate stiff area as the chuck is moved by the fingers prior to using the spanners. On some models, a tommy bar is used to lock the motor spindle.

Plunge action

Although the plunge action router was first introduced around 1950, it has taken a long time to gain the popularity it now enjoys, largely because of slow development allied with the economics of manufacture. The non-plunging router is now very much the exception, such are the advantages of the plunge action, especially for operations such as mortising, double-stopped grooving and similar cuts, and freehand decorative work. The extent of the plunge is another aspect of its specification and therefore its working capacity, and usually ranges between 45mm and 65mm ($1\frac{3}{4}$in and $2\frac{1}{2}$in). The body is usually mounted on twin bars secured to the base, the body sliding on the bars with springs that raise it when the

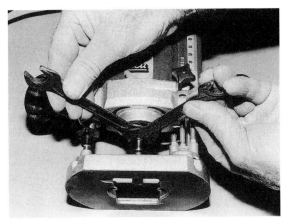

216 *Slackening the collet chuck*

217 *Small model from the Bosch range*

clamp is released. The body can be locked in any position along the bars, usually by turning one of the two handles provided. The alternative to this is a clamping lever conveniently positioned alongside one of the handles.

Movement of the body up and down the bars is also the way in which the projection of the cutter below the base is controlled, even when plunge action is not required. Projection of the cutter governs the depth of cut. All routers include a scale where the depth of cut can be pre-set, and some models also have available as an extra a fine-adjustment device whereby the depth can be very accurately set and adjusted. Another feature widely adopted on plunge action routers is the three-way turret depth stop. The turret incorporates three screws of different lengths, which are all adjustable, and as the turret is rotated and 'clicked' into one of its three locations, one of the screws is aligned with the end of the depth control. Thus the cutter can be set to cut up to three predetermined depths, a useful feature when similar cuts need to be made but to different depths – stopped and haunched mortises, for instance.

Removable bodies

A couple of routers on the market, small models produced by Bosch and Black and Decker, have motor bodies that can be detached. They have 43mm collars by which they are secured to the carriage, which slides on the rods, and thus they can be used in a different mode to most other routers. Because this collar size equals that of most drills, these two routers can be secured to drill stands, and thus used for overhead work.

219 *This Bosch router has a removable body*

Guide fences

An essential part of the router is the guide fence, and a straight fence is included with all routers. The fence is

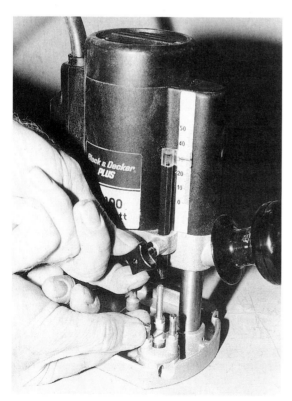

218 *Three-way turret depth control*

220 *Fine adjustment to fence on Elu router*

used when cuts parallel to a straight surface are being worked, the setting of the fence being controlled by a fine adjustment screw on many models. Often it is useful to have a fence that is longer than the one provided. It makes starting and ending many cuts easier. Fence extensions can be readily made by a strip of wood of appropriate size being screwed to the face of the fence, and holes are usually provided for this very purpose.

Roller guides

A straight fence is only of use for making straight cuts when working from a straight edge, and for curved edges a different arrangement is needed. This is provided by a roller guide, which with most models is designed to be secured to the straight fence when this is inverted within the base of the tool. The roller guide will operate on both concave and convex edges, or indeed those of compound shape.

Circle guides

Routers can be used for making circular cuts or cuts that are arcs of circles by using the circle guide. Essentially this is a pin, the end of which locates on the surface of the wood, or oddment of scrap secured to the surface of the wood. Either the guide pin is secured on its own arm to the router base, or fastened to the guide fence again when inverted. Thus the pin controls the movement of the router, allowing it to move only in a circular path.

223 *The circle guide in use with the Elu*

221 *Roller guide on the Hitachi*

222 *Forming groove on circular workpiece*

Guide bushes

A feature of the router is its ability to tackle irregular and complex shapes, and the ease with which it can follow templates and thus produce repetitive work to a high standard of accuracy. The roller guide already mentioned is designed for the roller to follow a ready-prepared surface, and cannot be used for internal cuts unless these are very large. For template work, a guide bush is required. This is a plate with a small cylinder in its centre. The plate is secured to the base of the router and the cutter protrudes through the cylinder. The shallow wall of the cylinder then follows the edge of the template which is attached to the workpiece, the cutter trimming off the waste to the edges of the wood to reproduce the outline of the template. Because of the difference in diameters between the cutter and the wall of the cylindrical part of the guide bush, the template is made smaller than the actual shape required to allow for this. A guide bush is provided with the router, which allows for basic template work to be undertaken, and additional guide bushes with different sizes of centre cylinder are available.

Capacity limitations

What has to be kept in mind at all times with any router making any kind of cut is that this must be carried out within the power limitations of the tool. While a high wattage is essential for the larger cutters, adequate power is just as important at all times, relative not just to the diameter of the cutter, but the amount of wood to be removed in making the cut. Often when making cuts with any size of router it is preferable to make two or more passes to achieve the profile required, and thus prevent overloading. Even with a powerful router it does not follow that all cuts can be made in one pass, as it also depends on the type and size of cutter being used. Cutting a deep groove, for instance, with a plain

224 *Bush guides are available in a wide range of sizes*

cutter of fairly small diameter could not be achieved in one pass, regardless of the power of the router. The strain on the shank of the cutter also plays an important role in the amount of material that can be removed in one pass, as hard dense wood offers far more resistance than most softwoods. The sharpness and condition of the cutter are also important, as with any cutters intended for wood, once they are partly blunt then they soon become very blunt as 'rubbing' takes over from 'cutting'. Blunt cutters also cause burning.

225 *Elu router with chippings extractor*

There are no precise rules that can be given regarding the relationship between router power, size of cutter, extent of cut, and rate of feed. The latter should be neither too fast nor too slow, but comfortable. Too fast a rate of working can result in a poor surface, and too slow can cause scorching even with a sharp cutter. Listen to the tone of the motor, and the drop in the rpm when under load. A drop of more than 25 per cent in the rpm is a sure indication of overloading.

Scorching

Scorching can also happen under certain conditions even with sharp cutters. The start of a plunge cut can, with certain cutters, cause burning, largely because not all the chippings can readily escape until the cut is enlarged. If with some cuts there is a tendency to scorch, this can be overcome by first removing the bulk of the waste, in more than one pass if need be, but leaving a final 1.5mm ($\frac{1}{16}$in) or thereabouts for a final, light cut to skim and clean up any scorching.

Chippings

Most routers have fan-cooled motors, with the exhaust air being expelled at the lower end. This helps considerably to keep the immediate cutting area clear of debris so that critical pencil lines can be seen. However, the combined action of the fan and the natural tendency of the cutters is to disperse the chippings far and wide. In common with so many other power tools, some manufacturers offer dust extraction kits that can be fitted to the router and the hose connected to a suitable chip collector. Some router tables also incorporate this option.

Cutters

The router is highly dependent on the cutters for much of its usefulness. While some cutters can carry out more than one form of profiling, such as a plain cutter forming a rebate or a groove, and certain moulding cutters being able to vary the extent of the moulding, generally speaking specific cutters are needed for specific cuts. While the range of cutters is almost endless, they do fall into certain types, quite apart from the size of the business end, and the diameter of the shank.

226 *Router cutters should be properly stored*

Plain cutters

Plain cutters cannot guide themselves, and the commonest way of using these is with the guide fence, roller guide, or bush guide. It is true that some plain cutters can be used with the router controlled in a freehand manner usually for lettering and other relief work. Most of these cutters will cut on their ends, that is have 'bottom cut', and are therefore suitable for plunge action modes of working as well as normal through and stopped cuts. The majority have two flutes or cutting edges, but most of those of small diameter have a single flute because of their size.

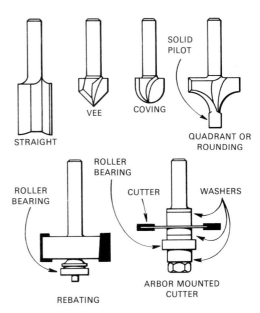

227 *Various types of router cutter*

Roller pilots

Some cutters incorporate a roller bearing at their lower end. This bearing follows the edge of the wood as the cut is being made, and thus guides the cutter in place of the fence. These cutters are therefore particularly suitable for shaped work, but usually the 'width' dimension of the cut cannot be adjusted, although in some cases an alternative diameter of bearing can be fitted to give some measure of adjustment. The depth of the cut can usually be varied by the normal adjustments of the router.

Solid pilots

A cheaper variation of the above is where the roller bearing is replaced by a solid cylindrical end to the cutter. This is known as the guide pin or pilot, and serves exactly the same purpose as the bearing. Because of the friction between the pin and the wood, scorching can take place fairly easily, especially when the side pressure is a little excessive, and the feed rate unnecessary slow. One advantage of the pin is that its smaller diameter enables it to reach further into acute angles than the roller bearing.

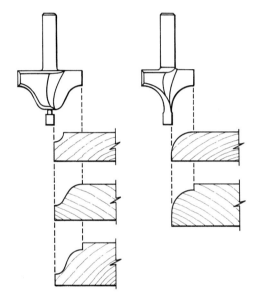

228 *Different profiles produced by varying height of cutter*

Arbors

Another variation is the arbor to which special cutters can be mounted, often more than one at a time so that compound cutting can take place. Arbors are particularly useful for mounting grooving cutters, which operate as tiny saws, usually with either two or three 'wings' or teeth. As well as being used for forming grooves, matching cutters can be obtained so that, for example, perfect tongue and groove joints can be formed.

Profile-scribers

A cutter development of more recent times is the profile-scriber set. Although essentially similar to the

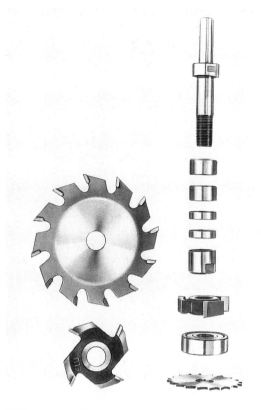

229 *Arbor and cutters*

arbor-mounted cutters mentioned above, the cutters are designed as matched pairs, and are intended particularly for panelled door and frame construction. One set of cutters forms the groove and the moulding;

230 *Matched moulding and scribing cutters*

the other cuts the stub tenon and scribed shoulder so as to create a perfectly fitting joint. Other cutters are produced to form raised panels, dovetails, T-slots and hinge recesses.

Tungsten carbide tipped (TCT) and high speed steel (HSS)

Most cutters are produced in both high speed steel, and in tungsten carbide. Those of HSS are suitable for short runs and light work, with the TCT being intended for heavier and more regular use. The latter are essential for use on man-made boards, and chipboard in particular. The blunting effect of chipboard is severe on router cutters, and sharpening is not easy. A sharpening jig is available that will sharpen cutters of both types, the jig fitting directly on to the router. With many cutters there is a loss of diameter when they are sharpened.

Rebating

One of the commenest cuts with a router is rebating. For this, a straight cutter is used, the diameter of the cutter chosen being greater than the width of the rebate. When making any kind of cut at the edge of the wood, the movement of the router must always be against the direction of rotation. This means that for

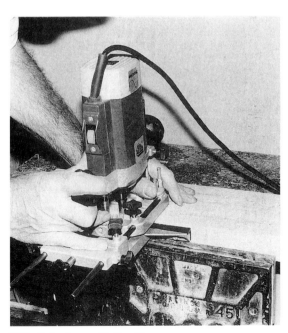

231 *Rebating with the Elu*

edge cuts, the router is moved from left to right, with the depth of cut taken at one pass as for all cuts being related to the width of the cut and the power of the tool. It is generally better to make successive passes by increasing the depth of cut – another use for the three-step turret – rather than by resetting the fence to increase the width of the cut.

End grain working

While the vast majority of router cuts made at the edge of the wood are formed along the grain, sometimes there is a need to make the cut across the end, or all around the workpiece as when making a panel.

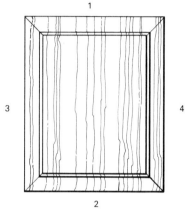

MAKE CROSS GRAIN CUTS FIRST

232 *Sequence of working when all edges require profiling*

Because end grain working is likely to cause some splintering at the corner at the completion of the cut, the end grain should always be tackled first. Any slight damage at the corner is then removed when the cuts along the grain are made. If cuts are to be made only across the grain, then scrap wood should be cramped to the workpiece so as to extend the width. The scrap acts as a 'run-off' and therefore any splitting is in the scrap and not the workpiece.

Moulded edges

The methods of working for moulded edges are just the same as for rebating, and stopped or double-stopped cuts simply require pencil marks to indicate the limits. Starting and stopping the cut are a matter of vigilance and careful control.

Narrow workpieces

Rebates on the ends of narrow work require a different technique, because of the limited support offered by the fence. Where several pieces need to be cut, then they can be cramped together and tackled as if one piece. If only one or two pieces are involved, the problem can be overcome by cramping on scrap wood of sufficient width to provide adequate support to the fence.

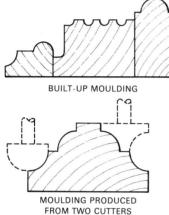

BUILT-UP MOULDING

MOULDING PRODUCED
FROM TWO CUTTERS

233 *Compound mouldings*

Grooving

In traditional hand work, a groove is a cut made along the grain, whereas a similar one made across the grain is called a trench. This is because of the different techniques that have to be used. In router work, though, a 'grooving' cutter works in all directions of the grain. The methods of forming the two cuts are similar, and because of this all cuts of this type are generally referred to as grooves. The cutters used for grooving are in fact straight cutters, available in a wide range of diameters. Where possible, grooves are best made to correspond to a standard size of cutter, although it is possible to widen a groove by taking more than one pass.

Grooves are normally made by controlling the router by the use of the fence, and for cuts that are away from the edges the direction of movement does not matter. Keeping close contact between the fence and the wood, though, is very important to produce a true cut. Where a groove that runs across the grain is fairly near the end the same technique is used, but only if the wood is sufficiently wide to provide proper support for the fence. Narrow pieces are tackled as for rebating the ends. For stopped grooves, it is generally better to commence the cut from the open end and

complete it at the stop line, while for double stopped grooves the plunge action is brought into use.

Guiding the router

Often, cross-grain grooves need to be formed well away from the ends and therefore beyond the reach of the fence. Some means of guiding the router is required, and the simplest method is by way of a home made guide, which in principle is similar to a T-square. This is secured to the wood by cramps some distance away from the cut required, with this spacing depending on the size of the base of the router. A

234 *Home-made guide for cross grain work*

slightly more sophisticated version has two arms, so that the router will just fit between without any slack. This ensures that the router cannot move off course, which is a possibility with the simpler guide.

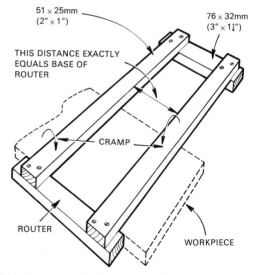

51 × 25mm
(2″ × 1″)

76 × 32mm
(3″ × 1¼″)

THIS DISTANCE EXACTLY
EQUALS BASE OF
ROUTER

CRAMP

ROUTER

WORKPIECE

235 *Router guide for cross grain working*

A specially designed tracking fence principally for cross-grain working is produced by Trend Machinery and Cutting Tools, and although initially developed for routers from the Elu range, a version is available that includes a universal sub-base by which almost any other router can be used. The essence of this jig is a duralium fence that is secured to the router, the fence then engaging in a groove in a 'slide board' made from polyethylene. The slide board is cramped to the workpiece, and thus controls the router along a set path. As well as being used for grooving, the tracking fence, or indeed any other form of guide, can, of course, be used for other cuts such as veeing and coving. In addition, the tracking fence can be secured to the wood at angles other than 90 degrees, and indeed can be used for cuts parallel to the grain. As well as the half-metre long slide board provided as standard, one double this length is also available. If a longer one is required, this can readily be made from 13mm ($\frac{1}{2}$in) thick ply, grooved in a similar way to the plastic original.

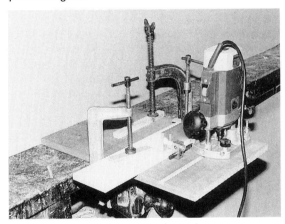

236 *Purpose-made tracking fence in use*

Another method of guiding the router for cross-grain work is by using the Wolfcraft Combinal, (see page 61). This combination tool is essentially an adjustable square, with its own clamp for securing to the work. As well as the 90 degree setting, other principal angles can be located automatically, and also any angle between zero and a right angle. When used as a router guide, it is worth the little extra trouble to secure the free end of the blade with a G-cramp.

When trenches are needed of the proportions used for half-lap joints, then it is very likely that the width will not coincide with the diameter of a standard straight cutter. Either the guide will have to be repositioned to enable the required width to be gained, or alternatively a guide made embracing two arms, wider apart than the base of the router, so that as

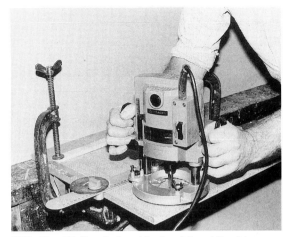

237 *The Wolfcraft Combinal guiding Hitachi router*

the router operates between these constraints the trench is cut to the width required. Trenches are best formed by tackling several members cramped together.

Boring holes

Straight cutters with bottom cut can be used for boring holes, especially the fairly shallow ones as required for dowel jointing. Where the dowel holes are located across the grain, then the various guides just mentioned can be used for ensuring the holes are square to the edge, and all in exact alignment.

Edge grooving

Often grooves need to be made in the edge of the wood, rather than the face. These can be formed either by using a straight cutter, or a special grooving cutter usually called a slotting tool. The latter are rather more efficient, faster cutting, and able to cut deeper at one pass. Because the cutter is mounted onto an arbor along with a roller bearing, the depth is predetermined and can only be altered by changing the cutter/bearing combination. The straight cutter is a little more suitable for stopped grooves, and when using this cutter the wood is held on edge. The slotting cutter because of its roller bearing will also function on curved edges, and in use the wood is secured horizontally.

Cutting mortises

From the point of view of a router cut, a mortise is essentially the same as a double-stopped groove. The router is ideal for small mortises, particularly stopped ones, because of the way they can be cut to a uniform depth with the bottom surface quite smooth, and within only a short distance of the reverse surface. Depth progression should only be around 6mm ($\frac{1}{4}$in) at a time. This reduces the strain on the cutter and prevents congestion from chippings. The technique can, of course, be used for through mortises. Mortises cut with a rotary cutter will always have rounded ends to the cut. Indeed, such mortises are often called slot mortises even when formed on stationary machines with rotary cutters. Despite the advice in much of the literature to round over the edges of the tenons to suit the mortises, most craftsmen prefer to square off the rounded ends of the mortise. This has to be carried out by hand chiselling.

Where a lot of mortising is carried out with a router, a second fence will be found to be a worthwhile extra piece of equipment, providing an additional fence can in fact be fitted on the opposite side of the base to the first. The advantage of working with two fences is that the router is fully controlled and cannot move out of line. The two-fence technique can also be used when edge grooving.

238 *Two fences help when mortising*

Templates

Shapes with completely curved or irregular edges can have these edges smoothed by using a combination of a template, bush guide and straight cutter. The template is made smaller than the size wanted for the item. This is because the cutter used must always be smaller than the guide bush used, the outside diameter of the small cylinder being the critical dimension.

239 *Guide bush in place on Elu router*

To arrive at the amount by which the template is made smaller than the product, the diameter of the cutter is subtracted from the diameter of the bush, and the result divided by two. For example, cutter diameter = 10mm, bush diameter = 14mm, then margin of reduction for template is:

$$\frac{(14-10)}{2} = 2\text{mm}$$

Note, though, that the template is made smaller where the outer edges of the template are the ones being followed. For an internal template, the aperture is made 2mm ($\frac{1}{12}$in) larger than the cut-out required on the item.

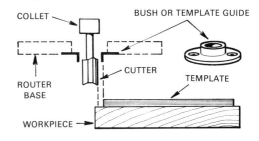

240 *Overhead template work*

For industrial applications where production runs can be very long, templates are best made in aluminium or tufnol. For general use, plywood is ideal. Birch ply 6mm ($\frac{1}{4}$in) thick is preferred. If thinner, the guide bush has insufficient surface on which to bear. If excessively thick, the 'reach' of the cutter becomes

241 *6mm (¼in) ply is ideal for templates*

unnecessarily long. The templates can be made by using the tools and techniques described throughout this book, and it is essential that the edges are well smoothed. Any inaccuracies on the template are reproduced on the wood. For most work, the wood is first cut to shape, but slightly oversize by no more than 1.5mm ($\frac{1}{16}$in) to allow for trimming. It is essential for the template to be secured to the wood for the routing stage, and this can be achieved in a number of ways. Panel pins can be used where the resultant holes are of no consequence. Cramps are suitable for certain shapes of template, or double-sided tape can be used.

The latter should be rubber based, enabling it to be peeled off the wood after use simply by thumb pressure, but it is only really suitable for use on smooth surfaces.

When trimming the edges, several passes might have to be made, depending on the thickness of the wood, how hard it is, and the power of the router. Movement of the router around the template is, as always, against the direction of rotation. This means when the router is hand-controlled, the movement on an external template is anti-clockwise, and clockwise where it is the internal shape of the template that is being followed. Where possible, template shapes with sharp internal corners should be avoided, as the router will always leave a radiused outline in these cases, as indeed they do for internal right angles. Where internal templates are in use, these are almost always for cuts made part way into the wood so as to produce a recess or sinking. Templates can be readily made and used in this way for handles, hinges and

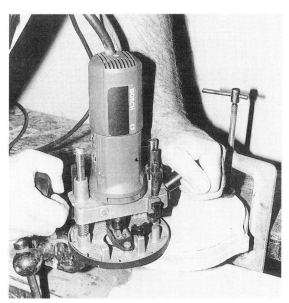

242 *Shield outline trimmed with aid of template*

243 *Template must be secured to workpiece*

244 *Typical internal template work*

and indeed the procedure is similar to that used for trimming purposes with the cuts being made progressively through the material. The method is not really suitable where a quantity of shapes need to be produced, and there is far more wear on the cutter compared with when the template and cutter are used for trimming purposes.

Inverted use

As well as being used in the handheld mode, where the wood is in the fixed position and the router moved, the router is particularly adaptable for use in the stationary mode. The usual way of using it in the fixed position is when inverted beneath a table, when the router effectively becomes a small 'spindle moulding machine'. Elu produce a combination bench which, as

locks, for instance, and therefore depth control of the cutter becomes critical. For this type of work, a small cutter is preferable as this produces a smaller area of uncut wood in what are usually 90 degree corners, which means less final trimming by chisel.

As an alternative to the method of cutting mortises already described, they can be formed by the use of an internally cut template. Where possible, the diameter of the cutter should equal the width of the mortise required, so that the template governs the length of the mortise and controls the movement of the router, normally along the grain.

It is possible to use a template when actually cutting the shape required. It is secured to the wood as before,

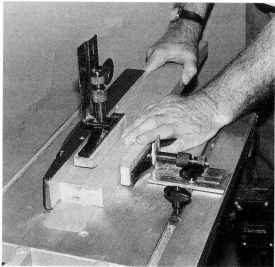

246 *The Elu combination table*

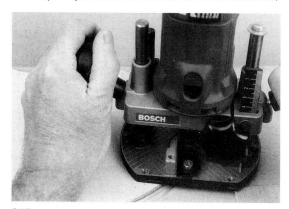

245 *Using template to cut outline required*

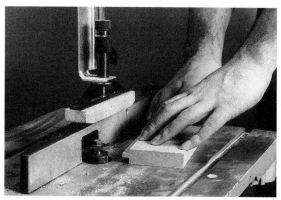

247 *Main fence and top pressure pad*

well as its use in conjunction with portable circular saws, will accommodate a wide range of routers from this manufacturer, and others if the user is willing to make adaptations. The table incorporates a no-volt release switch for added safety, as well as convenience because of the relative inaccessibility of the switch on the tool.

A principal part of the table is the fence, readily adjustable across the table and secured at one end only where there is a scale. The fence is of one piece, and there is no fine adjustment. A vertical pressure pad is provided with the table. A similar horizontal one is available as an accessory, along with a roller follower and a mitre fence with an adjustable face. Although a guard is not included, the pressure pads act as guards, and provide a good level of protection.

Many of the operations that can be performed in the handheld mode of working can also be carried out when the router is mounted beneath a table. Certain operations, though, lend themselves specifically to one of these modes of working. For example, tenons are best formed on the router table, while mortises are cut with the router handheld. With large and heavy workpieces, it is generally easier to carry out routing cuts with the wood in the fixed position, while the stationary mode is particularly advantageous with smaller pieces, and especially with those which would be difficult to hold properly when the router is being moved. The fine height adjuster is very useful when table routing.

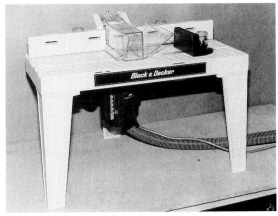

249 *The Black and Decker router table*

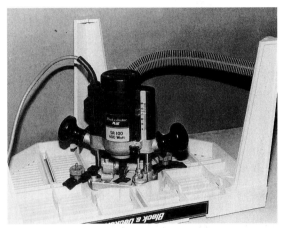

250 *Mounting of router beneath table*

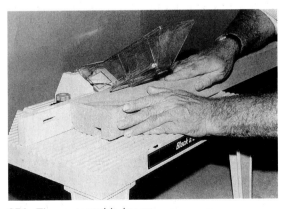

251 *The router table in use*

The Elu is a top-quality table, suitable for work of high professional standards, but there are others, including those aimed at more occasional use. Many are multi-purpose and are described in Chapter 11.

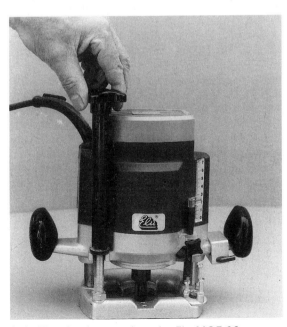

248 *Fine depth control on the Elu MOF 96*

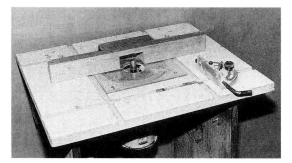

252 *Home-made router table*

It is not difficult to make a simple type of router table, and it can even include a cross-cut slide from another piece of equipment.

Straight work

Straightforward cuts such as grooves, rebates and mouldings, are normally made using the fence, assuming the edge of the wood is straight. Where the wood is quite wide, only one pressure pad can be used according to whether the material is passing over the table horizontally or vertically. Where possible, both pads are used, and especially when the wood is particularly small. Generally, where material of small section has to be profiled, it is far better where

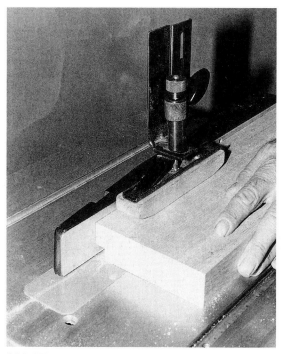

253 *Wide pieces require only one pressure pad*

applicable to commence with a fairly wide piece of material with its thickness equal to one of the dimensions of the section required, shape the edge as needed, then saw down to give the final dimension needed. Usually both edges of the wood can be worked on simultaneously to speed up the whole process. When the router is operating in the inverted position, the wood always passes the cutter from right to left.

Template work

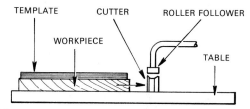

254 *Template work on router table*

Template work can also be carried out on the router table. The roller follower is positioned just over a straight cutter, which needs to protrude through the table very slightly more than the thickness of the workpiece. The template needs to be prepared to the identical size of the component. An allowance is not necessary as when using a guide bush but in fact a template either slightly larger or slightly smaller can still be used. The workpiece is cut marginally larger than the final outline required. If the template equals the actual shape needed, then the roller is positioned just above the cutter and level with the actual cutting edge. If the template varies in size from that required for the component, then the roller is laterally adjusted relative to the cutter by the same amount.

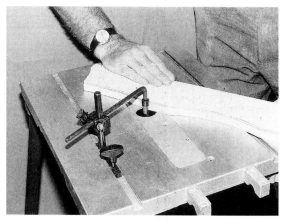

255 *Using plain cutter, roller follower, and template*

The template is secured to the top surface of the workpiece, then fed on to the roller when the cutter will remove the waste to the outline required. The lower corner of the template will be lightly removed with the cutter, hence the need to keep this as low as cutting will allow. It is important with this technique to ensure that all actual cutting is at the same point on the periphery of the cutter relative to the table or, because of likely differences in diameter between the roller and the cutter, the extent of the cutting could vary.

258 *The roller guide helps at the start of the cut*

Piloted cutter working

Both roller bearing and solid pilot cutters lend themselves well to use in the router table, where they are

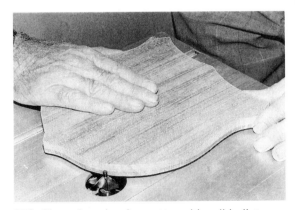

256 *Shaped work using cutter with solid pilot*

259 *This cutter has a roller bearing*

The roller follower can be used to provide this lead-in support, or a block of wood with rounded end cramped to the top of the table.

Routing assembled work

Frequently, the router is brought into use on a frame after it has been assembled, for instance in the case of a frame that carries a panel when it has been temporarily cramped together. In the case of an assembly such as a cabinet door, any moulding or similar cut to the outer edge can then be made across the ends of

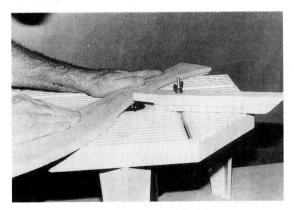

257 *Using wooden follower and plain cutter*

used without the fence for curved and shaped work. It helps to make the start of the cut both easier and safer if a lead-in fulcrum is provided a small distance to the right of the cutter. The wood is pivoted on this and slowly swung round so that the cutter begins cutting right at the end, or at the 'stop' line as the case may be.

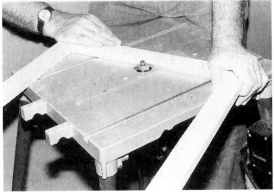

260 *Rebating a frame after assembly*

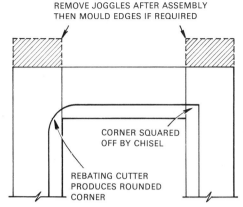

261 *Routing cuts after assembly*

the stiles once these have been levelled off. This would not be practicable before assembly. Assuming the edges are straight, this work can be carried out by using the fence. The inner edges of the frame can also be tackled on the router table, but it is necessary to see a self-guiding cutter. All cuts made into an internal right angle will produce a rounded corner. In the case of mouldings this has become quite fashionable and gives a sense of continuity to the profile. Where a rebate is made in this way, though, it is likely that a square corner will be required, and therefore completion of the corner will have to be carried out by chisel.

Cutting joints

When cutting tenons, the wood should first be cut to exact length, and a straight cutter of around 13mm ($\frac{1}{2}$in) mounted in the router. The height of the cutter above the table is set to equal the extent of the cheek of the tenon, and the fence positioned so that with the end of the wood fully against this, the cut is made alongside the shoulder of the joint. The mitre fence is

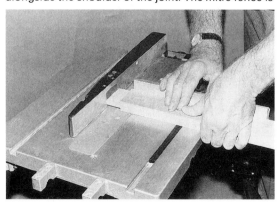

262 *Tenons formed with router*

used to support the wood as a series of passes across the cutter removes the waste. It is wise to mark out the first tenon fully to ensure that all the adjustments are correct, when all tenons subsequently cut should be identical. It is also wise to cut the mortises first, as it is easier to make small adjustments to the thickness of the tenons than vice versa. Half-lap joints at the end of the material are cut in a similar way, as are rebates so as to form tongues.

Small multi-purpose table

A small table is also produced by Elu specially for their MOF router. The fence is non-adjustable, but there is limited movement of the router beneath the table,

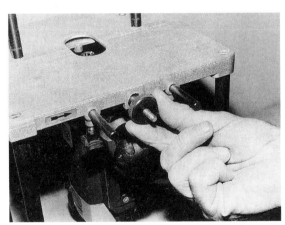

263 *The Elu accessory table in use*

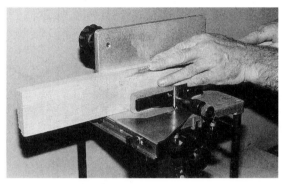

264 *Adjusting position of router beneath table*

which provides for adjusting the cutter relative to the wood, but cuts well away from the edge cannot be made. One pressure pad is provided. This will fit either horizontally or vertically, and a second one can be obtained as an accessory. Because the fence is very much higher than the usual pattern, this unit is

265 *One pressure pad is provided with this accessory*

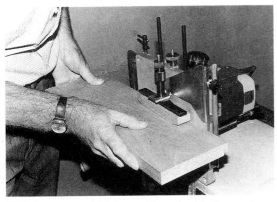

266 *Accessory table repositioned with the router horizontal*

particularly useful where the workpiece needs to pass over the cutters when standing on its end – forming the mating ends to slot dovetails, for instance.

This small table can be used with the main table horizontal when it stands on a set of legs, or with the table vertical and the fence horizontal when the two clamps provided secure it to a suitable bench. When horizontal, the table can be used for shaped work

267 *Shaped work on the accessory table*

with piloted cutters and a roller follower is included along with a shaped wooden block that provides the lead-in.

Because of the small and compact size of the table, it is possible to mount router, table and fence together and then use them in the mobile mode. The table and fence provide far greater support than do the router base and standard fence, and it is also possible to use this combination with the router horizontal.

268 *Accessory table in mobile use*

Overhead use

When a router is secured to a drill stand for use in the overhead mode of working most operations that can be performed on a router table can be carried out. There is the added advantage that the cut can be seen as it is being made, and work can be completed with or without the fence. If cuts need to be made in conjunction with a mitre fence this will have to be home-made. The golden rule of direction of working must still be followed, and for overhead work, the wood must move past the cutter from left to right.

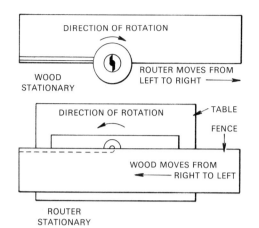

269 *Direction of movement*

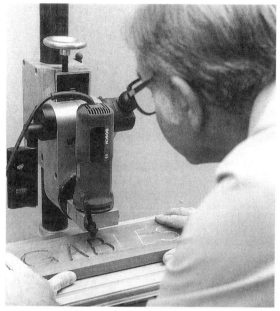

270 *Freehand work beneath fixed router*

271 *Angle cuts can be made on the Wolfcraft machining centre*

272 *Forming a mortise on the 5005*

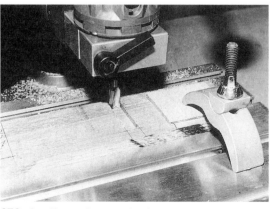

273 *Small trenches can be formed with precision*

Freehand surface decorative work including lettering can be carried out by this method of working.

The Wolfcraft 5005 machining centre allows for a wider range of operations than the usual drill stand, and indeed certain router functions can be performed on this piece of equipment that cannot readily be completed by other means. Because the router can be tilted out of vertical, cuts can be made other than the usual 'square-on' to the face or edge. Thus an angled groove can be cut, and a rebate with one or both its surfaces bevelled presents no problems. In addition, small workpieces can be clamped to the table of this machining centre, and movement controlled to very fine limits by the two hand wheels. The electronic scales instantly show the amount of movement being made, and the precision of the work that can be carried out is extremely exact. For stopped trenches and grooves, mortises, tongues, and other joints on small scale work, the Wolfcraft 5005 is an excellent work centre.

What has to be kept in mind is that drill stands, including the Wolfcraft 5005, will only accept drills and routers that have 43mm collars. Specialist suppliers of routing accessories do offer the equipment necessary to mount certain other routers onto stands, thus enabling overhead operations to be performed.

274 *Specially adapted drill stand to carry router*

Dovetailing jig

Several firms offer a dovetailing jig specially for use with a router, and this is much more sophisticated than the basic types of dovetailing jig. The pins and sockets are of equal size, but the jigs allow for the inner surface of the tails to be rounded over. Thus a

275 *The Elu dovetailing jig in use*

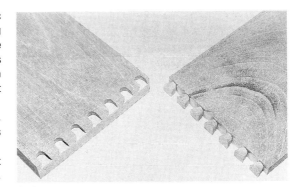

276 *The Elu jig cuts fully fitting joints*

fully fitting joint is produced as these rounded tails bed properly in the correspondingly rounded lower ends of the waste between the pins. The jigs allow wood up to 300mm (11¾in) wide to be jointed, and thicknesses between 12mm (½in) and 30mm (1³⁄₁₆in) can be secured between the clamps. With these jigs, it is also possible to cut the pins within a rebated end, as would be needed if a drawer were being constructed where the front was intended to overlap the carcase. A special guide bush and dovetail cutter are included with the kit.

Freehand working

As well as freehand decorative work when the router is fixed overhead and the wood is moved, freehand work can also be carried out in the more usual way whereby the router is moved over the wood. Practice is required before competence is gained, and with experience, the depth of the cut can be adjusted as cutting proceeds. Freehand work is best confined to small cutters, vee and round end cutters being particularly suitable.

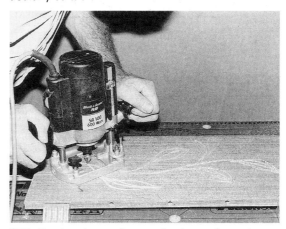

277 *Freehand use of router for decorative cuts*

Other aids to router work

In addition to lettering being carried out in a freehand manner, pantagraphs are available for use with routers. As well as being able to carry out lettering by following letter templates, they can also carry out copy routing, and cut circles and rectangles.

Other jigs produced include one for staircase housing, and one for forming the joints where kitchen worktops meet at an angle. The staircase jig requires the use of a heavy-duty router, and is designed so that it can be reversed to create an identical pair of strings. The jig is very adaptable so that different combinations of sizes of tread and riser can be catered for, as well as different thicknesses of wood. A special cutter is provided which cuts the housings with walls of 95 degrees, thus ensuring a tight fit between the components when the securing wedges are driven in.

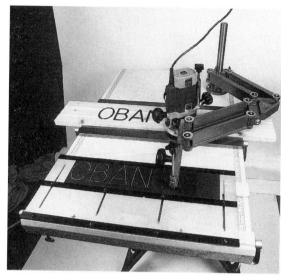

278 *Routergraph allows for precise copying*

279 *Special jig for kitchen worktops*

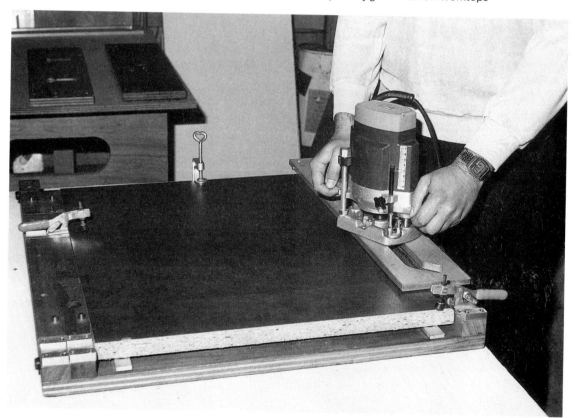

11 || Work Centres

Although all portable power tools are designed primarily for use in the handheld mode of working, many of them, as has already been described, also lend themselves to use when stationary and with the wood being moved. This method of use does not just offer an alternative way of carrying out a function, but usually provides scope for performing additional operations.

Stationary use of power tools has led to the development of the concept of the work centre. In essence, the work centre allows a number of portable tools to be mounted and used in the stationary mode, and thus enables the tool to perform more as a small machine. The facilities of the work centre, such as fence and cross cut slide, allow a wider range of operations to be carried out than when used handheld, with the table part of the centre giving good support and therefore enabling large pieces of wood to be handled with a high degree of safety. Generally, work centres cater for those tools that do not lend themselves to being simply inverted and cramped or held to the bench, although the range of tools that can be accommodated varies from centre to centre.

Work centres are generally designed to accept a number of models of power tools, of which circular saws, routers and jig saws are probably the most common. Because of the wide range of sizes of power tools and the varying features on them, especially relating to their sole plates, accommodation in the work centres is usually restricted to specific models from certain manufacturers. However, it is possible that models not specifically listed as being compatible will in fact fit, or the mounting brackets modified to allow for fitting to take place. It is essential, though, to ensure that the work centre and the power tool are properly matched, particularly relating to the positive securing of the tool. Work centre makers usually put a limit on the maximum size of tool that can be used with the centre. Work centres do not increase the capacity of the power tool – in fact, in some cases, this is reduced slightly – nor do they in any way make a poor tool into a good one.

Work centres vary considerably in the way in which they contribute to the manner in which the power tools may be used. Some, in fact, hold the wood in a fixed position while the tool is moved on to this, with the movement being in a controlled way. Many of these units are available in a basic form offering essential provision, with supplementary components being available to add to the scope.

The Meritcraft Powermate

The Powermate has a cast aluminium table with short folding steel legs. It can be used in this way as a bench version, or mounted on a special leg stand which raises it to normal working height. The switch unit is comprehensive and incorporates a socket into which the power tool is plugged, with the on/off switch well guarded. This unit can be locked into any position on any edge of the table, all the edges having provision for this in the casting. This also allows for the main fence to be similarly locked on to any edge, although when sawing only one edge is suitable.

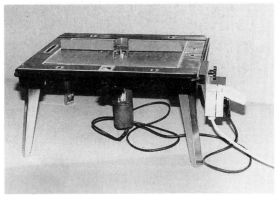

280 *The Meritcraft work centre set up for router use*

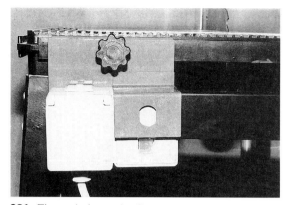

281 *The switch can be fitted to any edge of table*

282 *The Meritcraft takes a wide range of tools*

283 *Lower shelf provides storage for mounted tools*

The power tools themselves are mounted onto plated steel inserts that clip into the aperture in the middle of the table. A separate insert is used for each tool. This means that, when changing from one function to another, only the plate needs to be lifted out. The tool can remain fixed to this although there is a little overlap of usage of the plates relating to the sander. The lower part of the main leg stand is designed to hold a couple of the inserts, complete with tools, when not in use.

Work centres are particularly useful with a circular saw, and the Powermate has most of the facilities of a fixed saw bench. Much of what has already been said in Chapter 6 regarding the use of circular saws in a saw table applies to this work centre when used in this mode. However, the tilting facility of the saw cannot be used because of the restricted size of the gap through which the blade penetrates. Bevels of 45 degrees can be cut providing the special attachment is used. This is a cradle, preset to 45 degrees, that straddles the blade and supports the wood as the cut is being made.

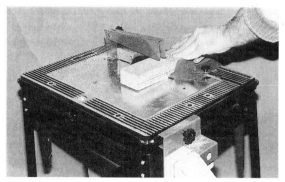

284 *Ripping on the Meritcraft*

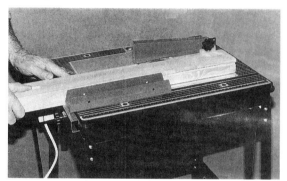

285 *The mitre fence fits to the right of the blade*

The router is considered by many to be the most useful of all the power tools, and it lends itself particularly well to stationary use as described in the chapter on this tool. With a cutter combining a pilot, all that is needed is for the cutter to protrude through the hole. Otherwise the router fence has to be used. The one provided with the Powermate can, of course, be adjusted to any position on the table, but the faces of the fence cannot be independently adjusted. This

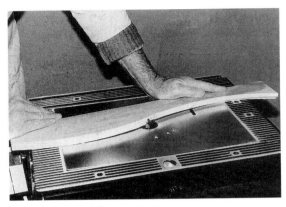

286 *Shaped router work*

imposes only a slight limitation on its use. However, it is also possible to use the rip fence from the saw set-up in conjunction with the router fence. By positioning the rip fence parallel to the one provided for the router and with the distance between them equalling the width of the workpiece, the wood can be fed between the two. This gives good control of the wood, and adds to the safety. A cover guard to the cutter is not provided as a part of the fence.

The jig saw used handheld is particularly suitable for work of fairly large dimensions, but this tool has its limitations when the wood is quite small, and especially where this requires intricate cutting. The problem relates to holding the work, rather than the actual cutting. Using the jig saw when inverted, and with the benefit of a good-sized table, is particularly helpful when small and awkwardly shaped workpieces have to be cut. On the other hand, this set-up is less suitable for large pieces of wood.

The use of the jig saw when inverted has another advantage. When used handheld, much of the sawdust from the cutting is deposited on the surface of the wood, often obliterating the line being followed. This does not happen when used stationary, as effectively the direction of cutting is reversed with cutting being on the downstroke. It does, though, raise a question about the 'face' of the work because of the risk of surface splintering. Handheld, any splintering of the fibres is on the upper surface, which is also where the lines to be followed will have been made. Used inverted, any splintering is on the underside, that is, the opposite side to the lines being followed.

Foam plastic is a material often used in conjunction with wood, especially for upholstery purposes. While straight cuts in foam can be made with a good knife, curved cuts in this material are far harder to make. The jig saw, when used inverted and with a knife blade

fitted, is an ideal way of cutting foam, and complex shapes are readily formed.

The jig saw plate can also be used to anchor belt sanders in the inverted position. However, use is restricted to a small number of models from the Black and Decker range. The top of the bodies of the compatible sanders are flat, with two threaded holes incorporated in this surface. All that is needed, therefore, is to screw the sander to the plate.

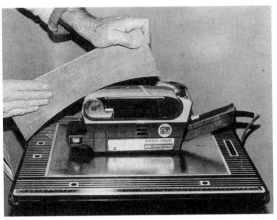

287 *Certain sanders can be mounted on the table*

An unusual feature of the Powermate is its ability to accept a drill stand. A hole in the table casting allows for the column of a drill stand to be secured directly to this unit, the column being locked in place with the knob provided. The hole allows a column of 25mm diameter to be positioned. Several of the Black and Decker stands have this diameter of column, and it is, of course, necessary to remove the base of the stand before mounting on the work centre.

The main advantage of using the drill stand as a part of the work centre is the large area of support offered to the material being bored, especially when compared with the inevitable small size of the base of the stand. It is therefore particularly useful when holes in large pieces of wood have to be made, boring hinge holes in doors, for instance. The stand can also be used as an overhead router, but such use is restricted to routers that can be detached from their sole plates and incorporate a 43mm collar.

The Triton

The Triton work centre is of Australian manufacture and is quite large and of heavy, sheet steel construction. It caters principally for the router, and circular and jig saws, and allows for circular sawing and routing to be carried out in two quite different modes of working. Either of these tools can be used when stationary, or the tools moved along the carriage with the wood remaining stationary.

288 *The Triton with extension table*

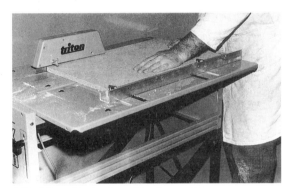

289 *The table in the wide rip position*

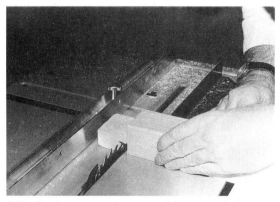

290 *Cutting tenons by the multi-pass technique*

To the basic bench unit can be added a leg stand, and also wheels to provide mobility. There is also a very large extension table that can be secured to the main unit. This extension is designed primarily for use when sawing, and provides a ripping width just over

1,220mm (48in) – large enough for a standard size sheet of ply or similar board to be sawn to any smaller size. The rip fence is of generous length, and there are two scales recessed into the top for the accurate setting of the fence parallel to the blade. The table to the main unit can be reversed end to end, thus allowing the blade of the saw to penetrate through a second slot. This is the wide rip position, when cuts up to 460mm (18in) in width can be made. The mitre fence has two faces. This provides for cuts of less than 45 degrees to be made, and by using both fences when preparing a mitre, a true combined angle of 90 degrees can be produced every time.

291 *A simple but generous sized mitre fence is provided*

The power tools are secured to a mounting plate that has nylon lugs at each corner. These lugs locate in aluminium channel rails mounted at the top of the unit under the table, and allow for the tool to be moved along the rails, and also provide for the mounting plate with tool attached to be inverted almost instantly. When the tool is inverted and the table in its upper location, the mounting plate clips into the stationary position.

The table can be repositioned within the body of the unit, beneath the top guide rails. Its height can be adjusted, and so support the wood at the correct position relative to the tool and the cut being made. A special fence is provided for use in this, the docking mode, to support the wood when cuts across the grain are being made. Indeed, the docking mode lends itself particularly well to working across the grain, and material up to 610mm (24in) can be held and cut. The height of the extension table can be adjusted so as to be level with the standard table to give adequate support to long pieces of wood.

For docking mode use, the tool and mounting plate are simply reversed so that the tool is uppermost, with both free to slide along the guide rails. As well as the

292 *The saw in the docking mode*

293 *Bevel cross cutting is easily achieved*

height adjustment of the table, height control can also be gained from the normal adjustments of the power tool. These also allow for tilting the saw whereby bevel cutting can take place. Trenching of large material is easy by taking a series of passes, and tenons too can be cut either in the docking mode, or in the normal position of working.

With the tool in the inverted position, the router operates in a similar way to when in the Powermate, except for the fence. The router fence provided with the Triton is in two parts, and is secured to the main rip fence with a gap between the forward and rear halves. The gap straddles the cutter, and a plastic guard is provided. The outfeed part of the fence incorporates a couple of adjusting screws whereby it can be moved slightly forwards relative to the near fence. This is a desirable feature when the whole edge of the workpiece is being moulded, such as when it is being fully rounded over. This is because for these types of cuts a little of the original edge is inevitably lost, making the moulded edge not quite level with the original. To compensate for this slight step as the cut is being made, the faces of the two parts of the fence

need to be out of line by an amount equal to the step created during cutting.

The same facility of the adjustment of the fence can be put to good use for cleaning up the edges of veneered and melamine-faced chipboard. Sawing these boards always presents some measure of difficulty regarding the slight splintering when cross cutting veneered boards, and the surface chipping when sawing melamine-faced boards in any direction. If the boards are sawn slightly oversize, then the edges can be trimmed using a plain TCT cutter in the router. The set-up is essentially like a planer working on its edge, that is, the edge of the cutter is level with the rear fence, with the difference in projection between this and the forward fence equalling the amount of material removed.

Using the router in the docking mode is particularly suitable for trenching, especially when the wood is quite wide. It is equally useful whether the trenches are stopped or through. For repeated stopped work the movement of the router across the wood should be controlled by cramping a block of wood across the rails so as to limit the movement. It can also be an advantage to secure a stop block to the fence when several pieces have to be trenched at a fixed distance from the ends.

It is essential when routing in the docking mode to cramp the wood to the unit. Failure to carry this out is almost certain to result in an inaccurate cut, as the wood will tend to creep slightly sideways under the influence of the rotating cutter if unrestrained. This is not the case when using the circular saw, as the force from the blade acts downwards, but the router is rotating in a different plane to a saw blade, hence the tendency for horizontal movement.

Within limits, it is possible to make angle cuts in the docking mode, both with the circular saw and the router. The supporting fence to the table is, however, non-adjustable when mounted beneath the guide rails, and it would therefore be necessary to secure a temporary wooden fence in place for non-standard cuts of this type.

Using a jig saw when inverted is essentially the same whatever type of work centre it is being used in. However, it can only be used in the Triton when inverted and does not lend itself to operating in the docking mode.

The Dunlop Powerbase

The Dunlop Powerbase has a large table of medium density fibreboard, mounted on a folding metal leg stand. Like the Triton, this work centre has two modes

294 *The Dunlop Powerbase*

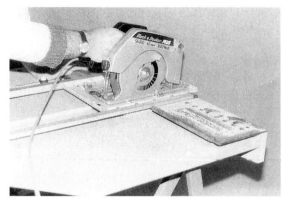

296 *For cross cutting, the blade height is adjusted*

of working, over or under the table. For overhead operation, the guide rails are secured above the table. These are adjustable in height according to the thickness of the material being worked on. Part of the guide rail system is a fence that aligns with the rear of the top, and against which the workpiece is held. Two clamps are provided with the Powerbase. These locate in any of the holes formed at 100mm (4in) centres in rows in the top. Using the clamps, the workpiece can be held against the fence.

as it becomes worn with use. The guide rails are preset at 90 degrees to the fence. This cannot be adjusted and cuts other than right angles cannot be made except by the introduction of packing. Work up to 65mm ($2\frac{1}{2}$in) thick can be accommodated under the guide rails, and cuts part way through the wood can be readily made by adjusting the body of the saw relative to its own sole plate. Thus trenches, cross-grain rebates, and tenons can be formed. A graduated arm that carries an adjustable stop is provided. This mounts on to one side of the guide system and is particularly useful for repetition work. Bevel cutting can also be carried out.

295 *Tools are secured to a steel mounting plate*

The tool is secured to the mounting plate by the usual arrangement of clips and screws, and the movement along the rails allows for cutting boards up to 610mm (24in) wide. Because the work is held directly onto the table, there is a groove down the centre of this to allow for the saw blade to cut through the work and penetrate into this space. Because the blade also has to saw through the fence in order to complete the cut, there is a renewable section for this

297 *Forming trench with router*

The mounting plate will also allow for a router to be secured in place, and in this mode is a particularly convenient way of forming trenches. With a plain TCT cutter in place, the ends of faced chipboard panels can be trimmed, the need for this having been explained in

298 *Using router without mounting plate*

299 *Using facilities of Powerbase when forming dowel holes*

300 *Provision is made for forming holes for concealed hinges*

relation to the Triton work centre on page 131. It is also possible to use the router without the mounting plate for certain operations simply by resting the sole plate directly on to the wood, and allowing the left-hand guide rail to guide the router, for instance, when trenching cuts are being made.

The guide rails and clamping system of the Powerbase provide a convenient way of holding the work and controlling a drill guide while holes are being prepared across the face of the wood. There is also a series of holes in the plastic wings that are part of the guide system, and these include two sets that enable blind holes to be made as required for con-cealed type cabinet hinges. These are made with the router with a guide plate fitted, and are formed at the exact distance required from the edge of the door for the hinge to be fitted, whether of the large or small pattern.

The circular saw on its mounting plate can also be used to convert the Powerbase into a saw bench. A removable panel in the centre of the table allows the saw to be dropped in place, and a simple fence, also of MDF, is provided. The clamps secure this to the top, and also provide for its adjustment. Parallel position-ing of the blade is established by measuring. Because of the way the riving knife and guard are secured in conjunction with the gap-plate immediately along-side the blade, bevel ripping cannot take place unless all these components are removed. This also applies to grooving and rebating. The Powerbase kit does not provide for cross cutting when the saw is in the fixed position, nor is separate switching included. This means that for the circular saw the handle has to be

301 *Ripping with saw in the fixed position*

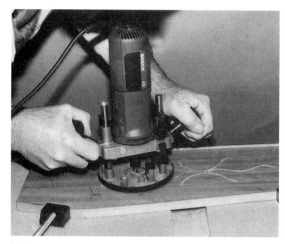

302 *Table top and clamps in use for holding workpieces*

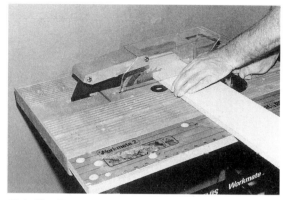

304 *The Black and Decker power tool table*

clipped in the 'on' position, and electrical switching of the tool controlled elsewhere.

Although the router can be secured on the mounting plate provided and secured in the inverted position in the centre of the table, a fence is not provided with the basic unit. Some cuts can, of course, be made by using the rip fence, and with a piloted cutter in the router a fence is not required.

Because of the layout of the holes in the top and the clamps that engage in these, the Powerbase provides a useful area on which pieces of wood can simply be held and gripped. Either a home-made fence needs to be made to replace the one that is a part of the guide rail system to help retain the wood gripped by the clamps, or alternatively the plastic studs from a Workmate could be used to complement the clamps. Thus material which, for example, requires sanding can be conveniently held, as can wood on which freehand routing needs to be carried out. Because of the way in which the clamps and the Black and Decker studs can rotate in the holes, work of irregular shape can be readily gripped. The top of this unit along with the clamps and studs can also be put to good use for certain assembly operations.

The Black and Decker Power Tool Table

The Power Tool Table is a Black and Decker product, designed for use with the circular saw, the jig saw, and the router. It is intended to be used on an optional folding leg stand, or alternatively in conjunction with a Workmate model WM 750. On the latter, it replaces one of the jaws and is gripped in position by the remaining jaw, which also extends the 740mm

(29¼in) by 460mm (18in) area of the table. The principal parts of the Power Tool Table are of tough plastic. The main table is of grid form, which allows much of the debris created when in use to fall through.

An unusual feature of this work centre is the use of templates to help support and retain the power tools. The templates have to be provided and prepared by the user; 6mm (¼in) ply is used for these and they are cut to rectangular shape so as to fit in the recessed space on the underside of the top. The templates also have to have a cut-out formed in them to correspond exactly with the sole plate of the particular tool being mounted. Holes are also needed in the ply, and by a combination of screws, plates and clips, the power tool is secured in the recess.

The fence is dual purpose, to be used both with the circular saw and the router. Secondary faces are provided to this fence. Both are adjustable, which is

303 *Fine adjustment is provided to the fence*

especially useful when routing, as already explained. A spring-loaded clear plastic guard fits between the two faces of the fence. It is always desirable to have the hole in the table through which the router cutter penetrates kept as small as possible, this measurement being consistent with the diameter of cutter being used. Because of this, a number of plastic inserts are provided. These have various sizes of centre hole and clip into the aperture in the table.

Two inserts are also included for clipping into the aperture in the table through which the saw blade protrudes. One of these has a narrow slot for normal square sawing. The other has a wider slot to allow the blade to be tilted over for bevel ripping. Ripping can take place either with both faces of the fence level, or with just the near half used by ensuring this is nearer the blade than the rear half. A riving knife and large guard are provided for the saw, and the design allows the riving knife to tilt when bevel cutting is taking place. A cross-cut fence is included, as is a clip for keeping the saw switch in the 'on' position.

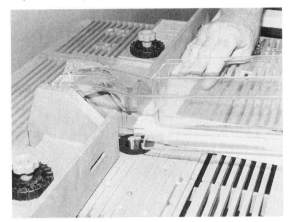

305 *The guard is of generous size*

306 *Two inserts are provided for the saw*

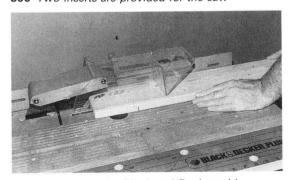

307 *Ripping on the Black and Decker table*

The Wolfcraft Variotec

The Variotec is made by Wolfcraft and caters for the three power tools common to the other work centres already described. The manufacturers claim that this unit will accept all makes of jig saw and router, and list a wide range of circular saws with which it can be used.

It is of galvanized steel construction. The special feature of this product is the swivelling top. This is

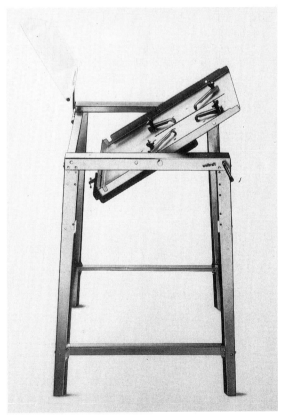

308 *The Wolfcraft Variotec*

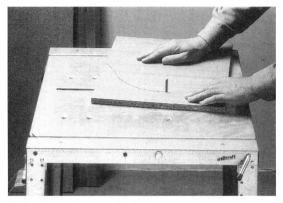

309 *Jig sawing on the Variotec*

620mm (24½in) by 565mm (22¼in) and is pivoted in the centre, thus allowing it to be flipped over. The tool being used is secured directly to the underside of the top, and therefore by inverting this, adding the tool becomes very straightforward. A rip fence is provided. This allows for ripping up to a maximum width of 310mm (12¼in). The riving knife of the saw is used, but a large clear guard is provided that is held by a support that fits on to the stand rather than the table top.

Various extras are available including outrigger work supports, a mitre fence, routing fence and independent plug and switch.

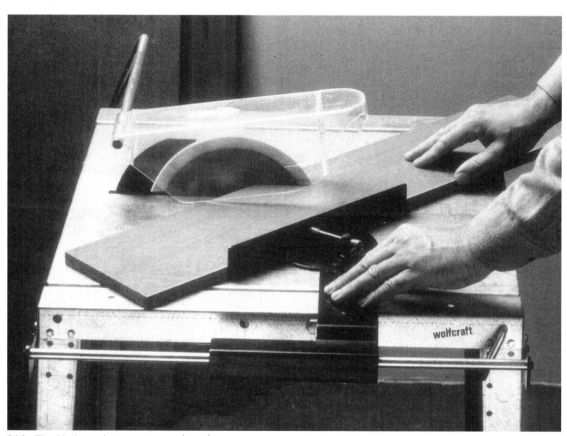

310 *The Variotec in use as a sawbench*

12 | Cordless Power Tools

Although cordless power tools have been in use since the 1960s, it is only in more recent times that they have gained ground. Indeed, more and more manufacturers are developing this type of tool, and many of them report cordless products as being the fastest growing area of their activities. The early tools of this type were the drills, but the range of cordless tools now available is considerable, and growing.

The obvious use for cordless tools is in situations where mains power is not available, but this is really only one aspect of their use. Often work needs to be carried out in places where it is simply more convenient to use a cordless tool than to lay on extension leads to the nearest power point – roof spaces, and in the foundation void, for example. Cordless tools are totally safe from an electrical point of view and therefore do not present the same hazard to the user when used in particularly wet conditions.

Nobody pretends that cordless tools have the same power as their mains-operated counterparts. The wattages are lower and, directly allied to this, the capacities are less. Generally, the speed of cordless tools is also less, although they score in other directions, for example the drills almost all have reversing facility and some have torque control. The big plus for these tools is the convenience they offer, a power tool that can be used in virtually any situation, even when mains is available, and complete freedom from leads and cables, plugs and sockets. They are most useful when regarded as being complementary to mains tools, not substitutes, and even in the workshop, experience indicates they are likely to be in regular use alongside the corded versions. Cordless tools are at their best when used handheld, rather than in stands and similar equipment. Indeed, the drills do not have a standard collar, and so cannot be gripped at this point.

311 *Small drill/driver from the Bosch range*

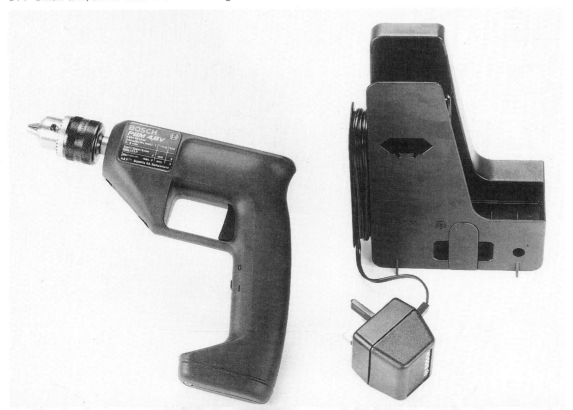

Batteries

The heart of the cordless tool is, of course, the battery, said to be rechargeable up to 1,000 times. Several nickel cadmium cells make up the power pack. The positive plates are of nickel impregnated with nickel salts. The negative plates are also of nickel but impregnated with cadmium salts and there is a nylon separator between. A chemical reaction takes place that produces the current, which theoretically is 1.25 volts per cell, but in practice is accepted as being 1.2 volts. When the battery is discharged oxidization takes place but no noticeable deterioration of the plates occurs during the chemical reaction, and charging then completely reverses what happens during use.

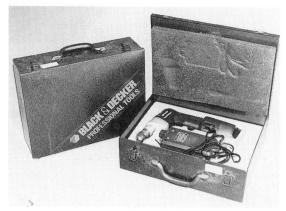

314 *Top professional cordless tools have their own kit boxes*

Most manufacturers offer these products across a wide range from the popular or amateur end of the market up to the professional, and a key difference between the extremes of the range is battery size, as measured by the number of cells in the pack. This varies from two cells up to nine, with four- and six-cell batteries being common with the lighter models, and six- to nine-cell packs preferred for the professional tools. Only with amateur models are the batteries built into the tool – the majority have the battery pack as a separate unit to the tool. With built-in cells, the charger is in the form of a plug. The lead from this is connected to the tool and around 16 hours are required for a full charge. The removable pack is inserted into the charger, most of which provide a full charge in one hour, although some from the middle of the range require three hours.

312 *Black and Decker professional battery being charged*

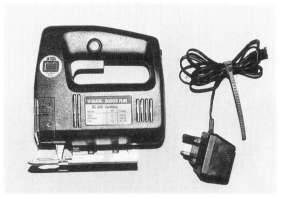

315 *The battery is charged via the plug*

When the batteries are both being charged and being used, heat is generated in the cells. This is quite normal and is a consequence of the chemical action. The tools, though, are at their best in average tempera-

313 *Compact model from Skil*

tures, and the third terminal in the batteries and chargers is in fact a thermal protector providing a cut-out should the battery temperature become too high. The batteries are able to deliver their full output up to the last 10 per cent of their charge, that is fall-off in performance of the tool only takes place when the batteries are approaching full discharge. The batteries give their best performance when fully exercised across their storage range. The batteries would take several months to discharge on their own without use, but should not be left in a discharged state.

316 *Professional model from the Makita range*

It is not wise to constantly 'top-up' a battery after only limited use and therefore partial discharge. The battery begins to act as if the partial use was actually its maximum capacity, a characteristic of the nickel cadmium cells known as the 'memory effect'. Unwitting use of the charger in this way will lead the user into believing that the whole tool is under-performing. However, the memory effect can be reversed by using the battery to a fully discharged state followed by a one hour charge, and repeating this cycle three or four times.

Battery packs, as might be expected, cannot be switched between different manufacturers' products, nor can six-cell and eight-cell packs be interchanged. However, the makers use the same design of pack in all their tools of a given cell rating, and batteries can therefore be interchanged within these limits. Battery packs can be bought separately, so the person who makes extensive use of these tools can have a reserve supply of energy to hand. On average, a top professional tool will give around 20 minutes of running time on a full charge. In most cases this is enough for several hours of actual work. Some manufacturers also produce a charger that will operate from a car battery by plugging into the cigarette lighter, an ideal solution when working well away from a mains

317 *Some models have wall brackets*

supply. It is claimed that a car battery in prime condition is capable of recharging an energy pack up to 12 times without serious effect.

Some of the products in the economy ranges include a wall holder for the tool. Returning the tool to the holder automatically makes an electrical connection so that charging takes place, providing the lead and charger are connected to the mains supply. With all types of chargers intended for power tools, over-charging cannot take place.

Screwdrivers

Among the smallest of the power tools are screwdrivers. They rotate at slow speed, will take alternative bit types, but do not have torque adjustment. The lighter

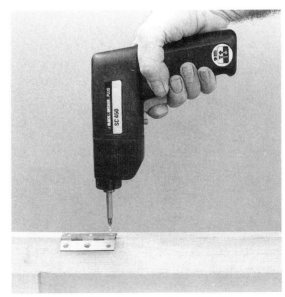

318 *Black and Decker screwdriver SC450*

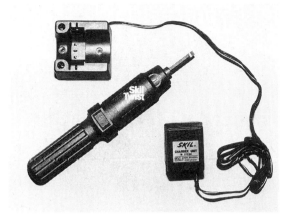

319 *Lightweight screwdriver from Skil*

ones from this end of the range will drive in screws up to 25mm (1in), or a little more in fairly soft wood and providing proper preparation has been made. With any kind of power screwdriving, cross-head screws are recommended rather than the traditional slot-head pattern.

Black and Decker drills

Black and Decker produce an extensive range of drills, including professional models of six and eight cells. Variations include two-speed models, hammer action, torque provision, and high and low single

320 *Black and Decker drill with battery and one-hour charger*

speed. Boring capacity in wood varies from 13mm ($\frac{1}{2}$in) up to 25mm (1in). This will depend on various factors including the type of bit used, and the hardness of the wood. While all these drills are capable of driving screws, a couple are designed specifically for both drilling and driving, and have a multiple setting clutch to prevent over-insertion of the screws. Maximum length of screw that can be driven

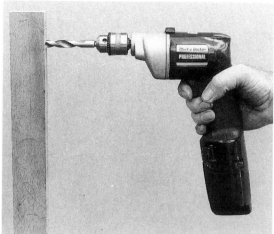

321 *Drills are at their best on smaller holes*

322 *Torque adjustment control on drill/driver*

is around 57mm (2$\frac{1}{4}$in). In common with the other manufacturers, these tools have provision for holding the chuck key within the body, and all Black and Decker professional drills are provided with a kit box, which houses the drill and the charger, with space for a spare battery.

Black and Decker jigsaws

Black and Decker offer two jigsaws; both are eight cell/9.6 volts. The 'Plus' model is very similar to the corresponding mains tool, and has an adjustable sole plate with provision for a rip fence. The battery is built into the tool, and the cutting capacity is 25mm (1in). The professional model is of very robust construction, and has the standard removable one-hour-charge

323 *Heavy-duty Black and Decker jig saw*

battery. The sole plate does not tilt, nor can it be fitted with a fence.

Makita drills

Makita produce a very comprehensive range of cordless products, embracing almost every type of power tool. Most are either six- or eight-cell, with the odd one of nine cells. The economy range of drills has built-in 7.2 volt batteries, with plug-in three-hour chargers. Variations across this range aimed at the occasional user include trigger-controlled variable speed, instant electric braking on release of the trigger, moulded finger grip adjacent to the trigger, and a built-in spirit level on the top of the body. The latter is especially useful for ensuring that the boring is taking place along a horizontal line.

324 *Makita drills with built-in batteries*

325 *A top model from the Makita range*

Different top models from this manufacturer also incorporate various features including dual speed, dual speed from zero controlled on the trigger, torque setting, instant braking, and moulded finger grip. The torque drill is especially intended for both drilling and driving screws, although with care this can also be accomplished by others in the range. A special drill produced by Makita is the angle drill, with a single speed of 700 rpm. This model is designed particularly for use where space is very restricted, with a distance of around 108mm ($4\frac{1}{4}$in) from the tips of the chuck jaws to the top of the body. Like all cordless tools intended for drilling, the angle drill is reversible, and will also drive in screws.

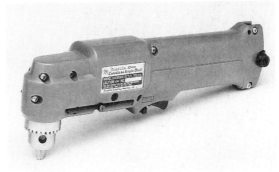

326 *Angle drill for confined spaces*

Makita saws

Three saws are featured in the Makita range. The jig saw will cut wood up to 25mm (1in), although half this thickness is a reasonable maximum in hardwood. It has an eight-cell battery pack, includes a safety lock on the switch, and has a tilting sole plate. The circular saw from the cordless range is the only one with a nine-cell battery, and again incorporates a device to prevent accidental starting. A spring-loaded guard, adjustable sole plate, and rip fence are all features of

327 *Professional jig saw from Makita*

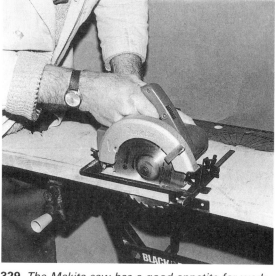

329 *The Makita saw has a good appetite for work*

328 *The Makita cordless circular saw*

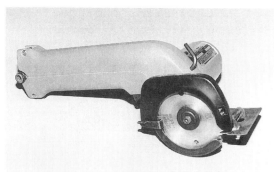

330 *The Makita buzz saw*

this tool. The TCT blade has a diameter of 160mm ($6\frac{1}{4}$in). This gives a maximum cutting depth of 55mm ($2\frac{1}{8}$in) at 90 degrees. The rpm of 1,000 is down on the desirable optimum, but this can be compensated for to some extent at least by the blade being particularly thin. This saw comes complete with a heavy-duty wooden box to house all the equipment provided as standard.

The small cordless saw is referred to as the buzz saw. It has a blade of 85mm ($3\frac{1}{8}$in), and a maximum depth of cut of 20mm ($\frac{3}{4}$in), which is adjustable. A

331 *The buzz saw is ideal for sheet material*

guard is provided, but bevel sawing cannot be undertaken. This saw is intended particularly for plywood and similar products, its fine teeth leaving a smooth surface. The small blade and light weight of this tool also make it suitable for making internal cut-outs.

Makita trimmer

A very unusual power tool to be found in the Makita cordless range is the trimmer. Resembling a router, it is light and compact, has a 6mm cottet chuck, and a speed of 8,000 rpm from its six cells. Its prime use is as a laminate trimmer, when the excess plastic needs to be trimmed flush at the edges. This cordless tool can carry out this operation to give a square edge, or a bevelled one, according to the type of cutter used. It

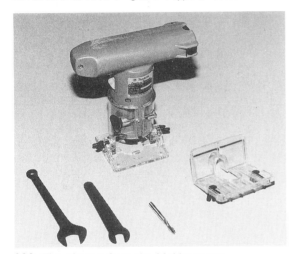

332 *The trimmer from the Makita range*

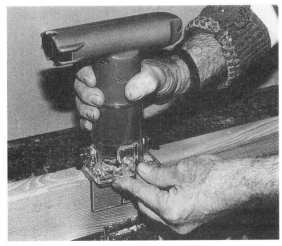

333 *Small rebates and grooves can be made with the trimmer*

will also carry out trimming work on plywood and similar sheet material used as a facing. The clear plastic fence is fully adjustable and readily allows progress to be checked.

The trimmer can also be used for forming small grooves and rebates, although only fairly light cuts can be made. In addition, it can tackle internal cut-outs, providing the material is not too thick. This little tool can also be used in a freehand manner, for recessed lettering, for instance, or for forming recesses as a part of a carved panel.

Makita sander

The Makita cordless sander is of one third sheet size, and is intended as a finishing sander for light use. It operates at 6,000 orbits per minute, and the abrasive paper is quickly and easily clipped into position.

334 *The Makita orbital sander*

All manufacturers of cordless tools have developed their designs to create products that are compact, surprisingly light in weight, and of sophisticated refinement. They perform well provided that their capacities are understood and they are used within their limitations. They offer an added dimension to power tools, especially in terms of their portability.

13 | Miscellaneous Tools

Tackers

Nails, tacks and staples are used far more in some branches of woodworking than others, but in any situation where they are used even to a moderate extent the power tacker is a great time saver. It is especially useful where wood and wood-based products are employed as cladding and wall panelling materials, and it has a particular application in upholstery work.

Most electric staplers are designed for driving in staples and nails up to around 30mm (1¼in), the maximum varying from model to model. In fact these power tools are at their best on staples, and the manufacturers offer a wider range of staples for their products than they do nails. It is essential to use only staples and nails made specially to match the tacker. To use the products of another manufacturer is almost certain to result in malfunction, most probably jamming the staples in the ejection gate. Both the staples and the nails for these tools are produced in linked cartridge form, but part-used cartridges can be removed and reinserted later enabling them to be used as required. These tackers have provision for visual checking of the quantity of staples remaining in the magazine.

Although power tackers generally are only able to cope with staples and nails of fairly short length, what has to be kept in mind is that when staples are used this is effectively the equivalant of driving in two nails of similar length. The grip effect of a nail is dependent of several factors including type of wood, length of nail, gauge of wire used, cross-sectional shape of the shank of the nail, and size and type of head. The staples used in these tools are usually produced from wire of square section. This normally gives a better grip than does a round section, and the nature of the 'head' of a staple adds considerably to its effectiveness providing it is visually acceptable. Nails produced for tackers are also of square section, with the heads of rectangular shape, which enables them to be

335 *The Bosch tacker*

produced in linked form. In many cases, where it is felt that even the longest nails or staples are the absolute minimum required for a particular application, additional securing benefit can be gained by using the staples at rather closer centres.

The staples for power tackers are produced in a range of lengths and in two popular widths of 4mm (3/16in) and 10mm (3/8in). They are also available in different gauges of wire, and some are produced with a resin-coated finish. The advantages of the staples with the legs 10mm (3/8in) apart are that the anchorage that they gain is well spaced, and the bridge between the legs offers good holding power when the material being stapled is quite thin, fabrics in particular. The disadvantage is that they are more noticeable, but this is often of no consequence.

Loading the staples is easy, with a spring-loaded bar moving them along the magazine and under the plunger. With some models, different magazines are required for certain sizes of staple and additional

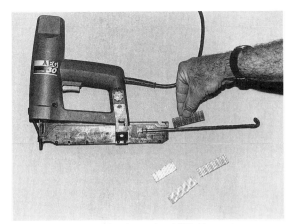

336 *Loading staples into the AEG Powertac*

components are available from a number of manufacturers including a device for stapling mitred components, and a guide that facilitates stapling along a line. It is also possible to obtain fittings in the form of clips and clamps that enable tongue and groove boarding to be secured by concealed stapling.

Normally, the tacker works on the basis of one pull on the trigger causing a single blow to be delivered via a plunger to the head of the staple or nail to drive this fully home. Because of the varying lengths of the staples used, and differing resistance of the wood, the strength of the blow can be adjusted so that the head of the staple or nail finishes flush with the surface, or otherwise as required. This adjustment is readily made by rotating the power control wheel, and all these tackers incorporate a basic safety device to prevent improper firing, which clearly could be highly dangerous. Before the tacker can be operated, the business end must be placed on the wood and depressed on its spring, which releases a safety catch.

With some models, it is possible to deliver a second blow to the staple, but only providing the tool is not moved between the first and second activating of the trigger. Lifting the tacker automatically brings the next staple beneath the plunger. On some models it is possible to set the tool so that double staples are delivered on a single pull of the trigger. Again this is a feature of benefit when fabrics are being secured.

With most types of tacker, it is possible to operate the tool on a 'continuous' basis. For this mode of working, the trigger is kept depressed throughout the whole operation and the tacker placed in the required position and the tool pressed down so as to override the safety lock and activate the mechanism. The tacker is then lifted, moved to the next position, and pressed down again. Tackers are able to operate at the rate of 30 blows per minute, and most are designed for short

period use. The tool heats up in use and extended continuous running causes overheating. After around 15–20 minutes of use at the maximum rate, the tool should be allowed to cool. If the strike rate is less than the maximum, the period of operation can be extended accordingly.

Power tackers draw a very high burst of current when impelling, and generally require protection from a 10 amp fuse. Performance falls off considerably if electrical demand is not met, and one cause of loss of impact strength can be when extension leads are in use. Extension leads should be kept down to ten metres or less, and only heavy-duty cable used.

Power hammers

A rather different type of impact tool for driving in staples and nails, along with other uses, is the Wagner Power Hammer. Whereas the usual type of tacker drives in the staple or nail in a single, adjustable blow, the power hammer operates at the rate of 100 blows per second, and this cannot be adjusted. There is a similar pattern of safety lock to that incorporated in the tackers.

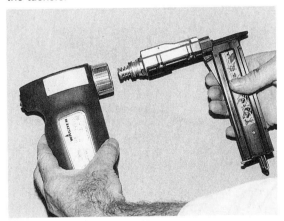

337 *The Wagner Power Hammer*

Various modes of operation are possible with this Wagner power tool. For use with staples, the magazine head is employed. This is essentially similar to the staple magazines of tackers. While it might appear that the continuous delivery of blows will result in the staples being driven in too far, in practice this does not happen because of the shrouding that surrounds the plunger. This tool can drive in staples up to 32mm (1¼in) in length. The staples must be supplied by the manufacturers of the tool and are available in two widths. While the magazine is capable of accepting both widths of staple, a different feed guide and plunger are required.

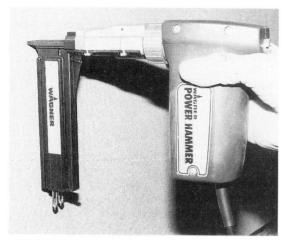

338 *Firing staples from magazine of the Wagner*

339 *Nails are driven in without the magazine*

Nails up to 60mm (2⅜in) can also be driven in with this tool but the method of operation for nails is quite different from that of staples. The magazine principle is not used. Specially prepared 'linked' nails are not required, and a variety of everyday nails can be driven in. A nail driving head is fitted to the tool to replace the head and magazine used when stapling, these heads being held to the body of the tool by a simple but very effective chuck that requires half a turn to provide a positive grip.

Two nailing heads are provided with the full kit – one for nails up to 30mm (1³⁄₁₆in) and the other for longer nails. Nails are individually hand fed into the tubular nozzle of the head where they are held against the end of the plunger by magnetism. Either the tip of

the nail if it protrudes from the head, or the end of the head itself if the nail is wholly within the nozzle, is then positioned on the wood and the trigger pressed. A rubber sleeve at the end of the nozzle acts as a cushion against the wood, and this also helps to control the extent of the driving relative to the surface. Various nails can be used providing the head of the nail will fit within the nozzle of the driving head. Longish nails of too fine a gauge will be found to have a tendency to bend when being driven into fairly hard wood, but this is no different to what would be expected if a hammer were being used.

A particular feature of this portable tool is that it can be converted to a power chisel. The special 22mm (⅞in) wide chisel blade is held by set screw into the blade head which, like the other attachments, is securely gripped in the chuck. Thus, with the trigger depressed, 100 blows per second turn the blade into a fast cutting chisel which is particularly suitable for constructional type woodworking. It cuts especially quickly and smoothly when working along the grain, and rather more slowly when cutting directly through the fibres of the grain. Because of this, the best result with the power chisel is probably achieved by combining its use with a saw where this is appropriate, for instance when cutting a trench. The limits of the trench are first cut by saw, and then the waste removed using the power chisel. An outside ground gouge is provided that can replace the chisel in the head, the sizes of the two being the same. These two

340 *The Wagner Power Hammer with chisel blade in use*

blades provide a useful means of removing the preliminary waste on a large piece of sculpture. A third blade, which likewise fits into the head, is available for cutting away putty when this has to be removed for replacing broken glass.

Bosch all-purpose saws

One of the younger generation of power tools is the all-purpose saw, of which Bosch produce two variations. It has a reciprocating action similar to a jig saw, but there the likeness ends. Both types made by Bosch have 550 watt motors, a maximum stroke rate of 2,600 per minute, and a stroke length of 26mm (1in). One has a fixed speed; the other has electronic control from 500 strokes up to the maximum.

Because this saw is designed to cut material with irregular and non-flat surfaces, the sole plate is small and is totally free to pivot. This means that, although the all-purpose saw will only make right-angled cuts, the angle of approach of the blade to the work can be varied simply by the way the saw is held against the workpiece. It is essential that the sole plate always be in close contact with the material, as otherwise proper control cannot be maintained and effective cutting will not take place. The sole plate can, however, be

moved and locked in a forward position. The wear on the blade is thus spread over a greater number of teeth, and this is particularly valuable when sawing fairly thin material.

342 *The all-purpose saw in use*

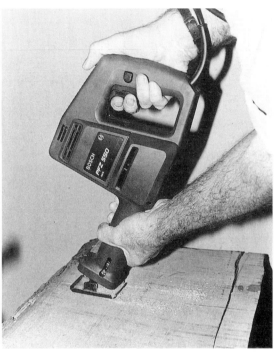

341 *The Bosch all-purpose saw with assorted blades*

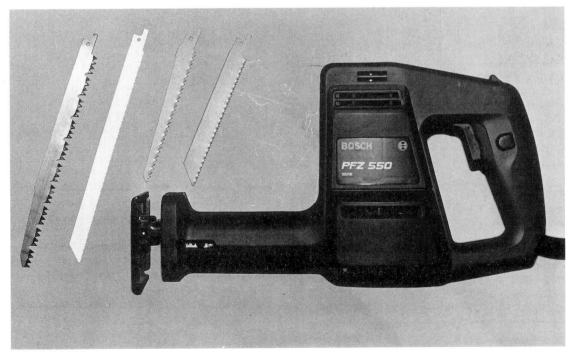

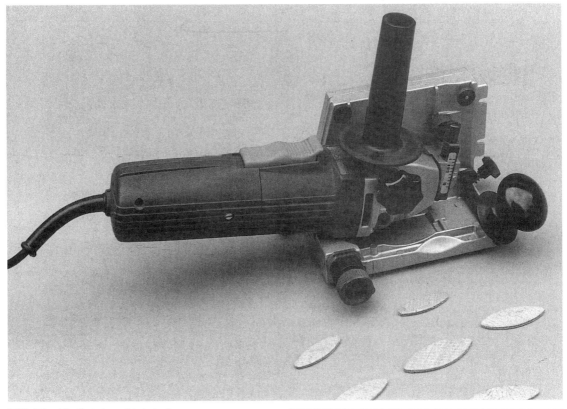

343 *The Elu flat dowel jointer/groover*

A single grub screw and clamp hold the blade in place. It can also be secured so that the teeth are pointing either downwards, or upwards. This adds to the ease and choice of holding the tool, especially when working in difficult-to-get-at situations.

Not surprisingly for a saw classed as all-purpose, there is a wide range of blades available for this tool. As well as fine and coarse for wood, another is intended for sawing wood with nails in it, also alloys and plasterboard. A blade with particularly coarse teeth is intended for small 'logging' work, and other preliminary sawing of large section timber, the 200mm (8in) long blade has a maximum cutting capacity of 150mm (6in). As well as an assortment of blades for cutting metals, there are others for plaster-board, building blocks, and plastics. One is particu-larly flexible because of its bi-metal construction, and this allows for a piece of material protruding through a surface to be sawn off flush. This blade is also ideal for cutting through metal tubing, and plastic rainwater piping. Because of the considerable cutting capacity of this tool, most of the blades are quite wide and these are primarily intended for straight rather than curved cutting.

ELU flat dowel jointer/groover

Traditional cylindrical dowelling has been around for a very long time indeed, but now we have an alternative to this old method of using dowels, known as flat dowels, or biscuit dowels. The dowels are specially made to suit this technique of jointing, and they cannot be satisfactorily produced even in a well-equipped workshop. Three sizes of dowels are manu-factured, although they are all 4mm ($\frac{1}{8}$in) thick, and they are of approximately elliptical outline. The sizes range from 45mm by 15mm ($1\frac{3}{4}$in by $\frac{5}{8}$in) to 60mm by 24mm ($2\frac{5}{8}$in by 1in). They are usually made of beech, and cut so that the grain runs diagonally to give the maximum strength in use. The special characteristics of these biscuit dowels lie in the fact that during manufacture they are compressed a little, and have their surfaces indented slightly. Thus when glued in place, the moisture from the glue is partially absorbed by the dowels, which swell a little as a result, and in addition, the indentations act as a key for the adhesive.

To form joints using the biscuit dowels, a special tool is required known as a flat dowel jointer. In essence, this tool is a circular saw. The blade has a diameter of 100mm (4in) and its TCT blade will give a maximum depth of cut of 24mm (1in). It is fitted with

344 *Typical biscuit dowel joint*

a fence, with the fine adjustment of the distance from the blade to the fence being controlled by a micro-screw that actually moves the sole plate on which the fence is mounted. Because the dowel joints are often used on boards of 16mm ($\frac{5}{8}$in), 19mm ($\frac{3}{4}$in) and 22mm ($\frac{7}{8}$in) thickness, there are specific markings on the body of the tool so that the fence-blade setting can be quickly made to the centre of the material of these common thicknesses. The depth of cut of the blade is also controlled by a micro-screw. Thus precise setting can be made and adjustments are aided by a graduated scale.

To form a recess for a flat dowel, a plunge action is required from the tool so as to make a cut that is segmental in outline. Adjustments therefore have to be made to the settings of the tool so that the extent of this cut corresponds to the size of dowel chosen for the particular job. In fact, the cut needs to be such that it is very slightly greater than half the size of the dowel, to allow for a little clearance and thus enable the components to be brought fully home. The plunge cut is made by holding the sole of the tool on the required position on the wood, then raising the handle, which is pivoted, and this lowers the blade up to the 'stop' setting. A line on the body of the tool is used to align the centre of the blade on the centre of the cut required. The thickness of the blade and the dowels correspond exactly to one another, and this, com-bined with the precise settings that can be made, leads to a very high degree of accuracy in forming joints. Proper preparation of the wood in the first place is, of course, essential, as is some marking out to indicate the positions of the dowels.

Clearly the fence has limitations as to the extent to which it can be set, and is intended for use when the cuts are on the edge of the material, or close to it.

345 *Forming joint on end of component*

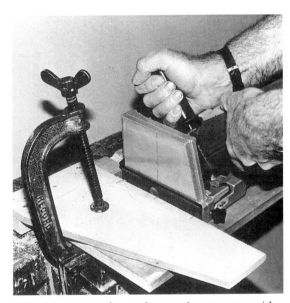

346 *Joints away from edge require separate guide*

Where cuts need to be made well away from the edge or end of the wood such as for shelves within a cabinet, a batten needs to be cramped to the wood to guide the tool. For cuts that have to be made in a mitre, a special 45 degree fence is included with the kit.

Although when used for cutting the recess required for a flat dowel the tool follows a plunge action, the jointer can be used in a forward manner so as to cut grooves along the length of the wood. Grooves wider

347 *Forming joint in mitre*

348 *Precise grooves can be formed with this Elu tool*

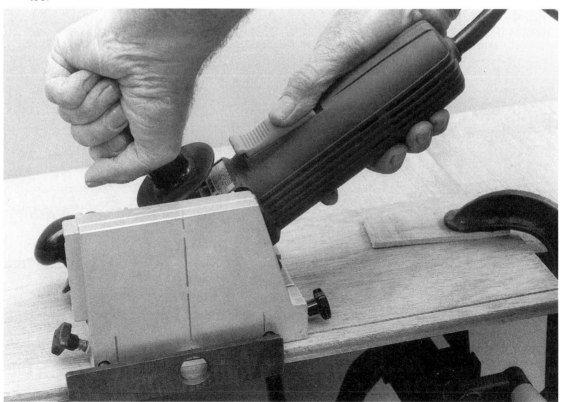

than the thickness of the blade can be made by taking multiple passes, and stopped cuts can be readily formed. Small rebates can also be cut, so tongue and groove joints are easily prepared with this tool. It is an extemely accurate tool partly because of the precise adjustments that can be made, and also because the small diameter of the blade allied with its thickness make it very rigid.

The flat dowel jointer is also especially useful for sawing veneered chipboard. The small, high revving blade combined with the special plastic insert that reduces the gap surrounding the blade to virtually nil enables cutting to take place with minimal disturbance of the surface fibres.

This Elu tool is suitable for forming flat dowel joints in all kinds of timber including man-made boards, and has a particular applications in cabinet-making of all types.

Glossary

Arbor A revolving spindle on which a blade or cutter is mounted

Bevel The edge of a piece of wood when planed at an angle other than a right angle

Bevel Cut A saw cut made at an angle other than a right angle to the face of a piece of wood; this may be along or across the grain

Box Comb A corner joint where fingers are formed on the ends of both members so as to interlock when fitted together

Carcase The main assembly of a cabinet when under construction

Chamfer The corner between two adjoining surfaces that is planed away; usually made at 45°

Collet Chuck A precision chuck used on routers whereby the cutters are held; the size of collet must match the size of shank of the cutter placed in it

Edge The two long narrow surfaces of a piece of wood

Electronic Speed Control The speed control of the tool gained through electronic circuitry whereby electrical power is delivered to the motor without loss of power at speeds below the maximum

Face The better of the two broad surfaces, or sides, of a piece of wood

Fence A guide of wood or metal enabling wood to be properly guided as it passes over or past a cutter or blade; alternatively it enables a tool to be guided as it moves along wood

Groove A recess formed in the wood when made along the grain (*See also* **Trench**)

Gullet The space between the teeth on a saw, particularly relating to a circular saw blade

Housing A joint formed when the end of one piece is fitted into a trench formed in the other

Lap A joint where one piece overlaps the other; also the joint when a strip of abrading cloth is formed into a belt by overlapping the two ends

Loose Tongue A thin strip of wood inserted into grooves in two pieces to join them together

Mitre Cut A saw cut made at an angle other than a right angle to the edges of a piece of wood; usually made at 45°

Pendulum Action The reciprocating action of jig saws when combined with a swinging forwards-backwards movement

Peripheral Speed Used for circular saws and measured in feet or metres per minute at which the teeth pass through the wood

Pin The part of a dovetail joint where the slope is on the end surface of the wood

Pistol Grip The usual arrangement of the handle of a drill where the handle is positioned at approximately 90° to the axis of rotation

Plunge Action The arrangement on a router whereby the body can slide vertically in relation to the sole plate

Rebate The corner of a piece of wood when cut away to form a step

Ripping Sawing along the grain

Scroll Action The action of jig saws whereby the blade can be made to rotate in a vertical plane as sawing takes place; this facilitates intricate cutting

Shank The upper part of a bit or cutter that is gripped in the chuck

Socket The part of a dovetail joint into which the pins are fitted

Sole Plate The flat, lower part of many tools, including routers, circular and jig saws, that is in contact with the wood

Speed Usually rated as the revolutions per minute (or rpm)

Surface Ripples The undulations running across the grain of the wood when a power plane has been moved too quickly over the wood

Tail The wood left when the **Socket** is cut for a dovetail joint

Tongue The projecting wood formed when two rebates are cut along the edge of the material; usually made to engage in a groove, and thus form a joint

Tooth Pitch The distance apart of teeth on a blade, measured from tip to tip

Trench A recess cut across the grain, distinguished from **Groove** by the direction of the grain

Trigger Action The speed control on a drill governed by the amount of depression of the on-off switch, which is in the form of a trigger

Wattage The usual measure of the power of a small motor, normally taken to be the input wattage; one horse power equals 760 watts of output rating

Suppliers

Britain

AEG (UK) Ltd
217 Bath Road
Slough
Berkshire
SL1 4AW
Tel. 0753 872101
Power tools and accessories

Black and Decker
Westpoint
The Grove
Slough
Berkshire
SL1 1QQ
Tel. 0753 74277
Power tools and accessories, saw blades and router cutters

Robert Bosch Ltd
PO Box 98
Broadwater Park
North Orbital Road
Denham
Uxbridge
Middlesex
UB9 5HJ
Tel. 0895 838383
Power tools and accessories, and jigsaw blades

Ceka Works Ltd
Pwllheli
Gwynedd
North Wales
LL53 5LH
Tel. 0758 612258
Drill stands and guides, boring bits for wood and drills for metal, and wood milling cutters

Cintride Ltd
Ashford Roadworks
Bakewell
Derbyshire
DE4 1GL
Metal and masonry drills, and TCT abrasive discs and sheets bonded to metal and flexible backings

Clico (Sheffield) Tooling Ltd
Unit 7
Fell Road Industrial Estate
Sheffield
S9 2AL
Tel. 0742 433007
Bits, countersinks, counterbores, plug cutters and router cutters

Elu Power Tools
Part of Black and Decker (*See above*)

English Abrasives Ltd
PO Box 85
Marsh Lane London
N17 0XA
Tel. 01 808 4545
Abrasives of every grade and grit produced in the form of sheets, discs, belts and bobbins, and safety products

Florin Ltd – Vitrex Tools
457–463 Caledonian Road
London
N7 9BB
Tel. 01 609 0011
Chuck keys, countersinks, rotary files, abrasive discs and safety products

Hitachi Power Tools (UK) Ltd
Precedent Drive
Rooksley
Milton Keynes
MK13 8PJ
Tel. 0908 660663
Power tools and accessories

A. Levermore and Co. Ltd
24 Endeavour Way
Wimbledon Park
London
SW19 8UH
Tel. 01 946 9882
Wood boring bits

Makita Electric (UK) Ltd
8 Finway
Dallow Road
Luton
Bedfordshire
LU1 1TR
Tel. 0582 455777
Power tools and accessories

Meritcraft Ltd
Martindale Industrial Estate
Hawkes Green
Cannock
Staffordshire
WS11 2XN
Tel. 05435 73462
Meritcraft work centre

MK Electric Ltd
Shrubbery Road
Edmonton
London
N9 0PB
Tel. 01 803 3344
Electrical safety equipment

Rabone Chesterman Ltd
Summer Hill Works
Camden Street
Birmingham
B1 3DB
Tel. 021 233 3300
Hole saws

Racal Safety Ltd
Beresford Avenue
Wembley
Middlesex
HA0 1QJ
Tel. 01 902 8887
Safety products

Record Marples Ltd
Parkway Works
Sheffield
S9 3BL
Tel. 0742 449066
Wood boring bits, router cutters, drill stands, square mortise chisels and bits

Skil (UK) Ltd
Fairacres Industrial Estate
Dedworth Road
Windsor
Berkshire
SL4 4LE
Tel. 0753 869525
Power tools and accessories

Trend Machinery and Cutting Tools Ltd
Unit N
Penfold Works
Imperial Way
Watford
Hertfordshire
WD4 4YF
Tel. 0923 249911
Comprehensive range of tools including router cutters, router jigs, aids, bits and countersinks

Triton Workcentres (UK)
PO Box 128
Bexhill-on-Sea
Sussex
TN40 2QT
Tel. 0424 216897
Triton workcentre

Wagner Spraytech (UK) Ltd
3 Haslemere Way
Tramway Industrial Estate
Banbury
Oxfordshire
OX16 8TY
Tel. 0295 65353
Power hammer, chisel and gouge

Wolfcraft
BriMarc
PO Box 100
Leamington Spa
Warwickshire
CV31 3LS
Tel. 0926 450370
Power tools, drill stands, machining centres and a wide range of accessories for power tools

America

Abbey Tools
616 North Brookhurst Street
Anaheim
CA 92801

Aviation/Industrial Supply
PO Box 38159
Denver
CO 80238

Aeg Power Tools Co.
New London
CT 06320

Black and Decker – ELU
10 North Park Drive
Hunt Valley
MD 21030

R. Bosch Power Tool Co.
One hundred Bosch Boulevard
Newburn
NC 28562-4097

Cintride Agency
Anglo American Enterprise Co.
401–403 Kennedy Boulevard
Somerdale
NJ 08083

Clico Agency
Woodcraft Supply Co.
41 Atlantic Avenue
Box 4000
Woburn
MA 01888
Also at:
Garrett Wade Co. Inc.
161 Avenue of the Americas
New York
NY 10013

Cascade Precision Tool Co. Inc.
PO Box 848
Mercer Island
WA 98040
Router cutters

Econ Abrasives
PO Box B867021
Plano
TX 75086
Abrasives

Freud
218 Field Avenue
High Point
NC 27264

Hitachi Power Tool (USA) Ltd
4487 East Park Drive
Norcross
GA 30093

Makita (USA) Inc.
12950 East Alondra Boulevard
Cerritos
CA 90701-2194

MLCS
PO Box 4053 F
Rydal
PA 19046

Racal Health and Safety Inc.
7309 Grove Road
Frederick
MD 21701

Santa Rosa Tools
1651 Piner Road
Santa Rosa
CA 95403

Severn Corners Ace Hardware Inc.
216 West 7th Street
St Paul
MN 55102

Skil Co
4801 Peterson Avenue
Chicago
IL 60646

Tool City
14136 East Furestone Boulevard
Santa Fe Springs
CA 90670

Whole Earth Access
2992 Seventh Street
Berkeley
CA 94710

Index